Preface

This Mini Atlas is a condensed version of the *Textbook of Biochemistry for Medical Students*. The original textbook is now running into 5th edition, and entering the 12th year of its existence. With humility, we may state that the medical community of India has warmly received the original textbook. In retrospect, it gives immense satisfaction to note that the textbook served the students and faculty for more than a decade.

There was a consistent demand from the students that a condensed version would be very useful for revision of the whole biochemistry; hence, this Atlas. Most of the figures, charts tables and boxes of the textbook have been reproduced in this book with minor modifications. Exhaustive explanations are given, so that the reader will be able to grasp the important points without any difficulty. This book will give a quick reference to the essentials of Biochemistry, and will be very useful to recollect the basic facts.

The remarkable success of the Textbook of Biochemistry was due to the active support of the publishers. This is to record our appreciation for the cooperation extended by Shri Jitendar P Vij and his associates of M/s Jaypee Brothers Medical Publishers (P) Ltd, New Delhi.

We hope that the students and faculty members will find this Atlas useful. Now this is in your hands to judge.

End of all knowledge must be building up of character
—Gandhiji

June, 2007

DM Vasudevan
Sreekumari S

Contents

Section 1

Chemical Basis
of Life

1. Subcellular Organelles

Cells contain various organised structures, collectively called as cell organelles. When the cell membrane is disrupted, the organised particles inside the cell are homogenised. They could then be separated by applying differential centrifugal forces.

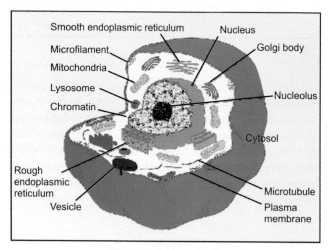

Fig. 1.1. A typical cell

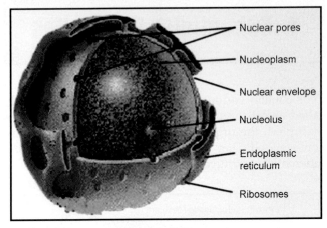

Fig. 1.2. Nucleus

Nucleus is the most prominent organelle of the cell. All cells contain nucleus, except mature RBCs in circulation. In some cells, nucleus occupies most of the available space, e.g. small lymphocytes and spermatozoa. Nucleus contains the **DNA**, the chemical basis of genes which governs all the functions of the cell. It is further organised into **chromosomes**. DNA replication and RNA synthesis (transcription) are taking place inside the nucleus.

Table 1.1. Metabolic functions of subcellular organelles

Nucleus	:	DNA replication, transcription
Endoplasmic reticulum	:	Biosynthesis of proteins, glycoproteins, lipoproteins, drug metabolism, ethanol oxidation, synthesis of cholesterol (partial).
Golgi body	:	Maturation of synthesised proteins.
Lysosome	:	Degradation of proteins, carbohydrates, lipids and nucleotides.
Mitochondria	:	Electron transport chain, ATP generation, TCA cycle, beta oxidation of fatty acids, ketone body production, urea synthesis (part), heme synthesis (part).
Cytosol	:	Protein synthesis, glycolysis, glycogen metabolism, HMP shunt pathway, transaminations, fatty acid synthesis, cholesterol synthesis (part), heme synthesis (part), urea synthesis (part), pyrimidine synthesis (part), purine synthesis.

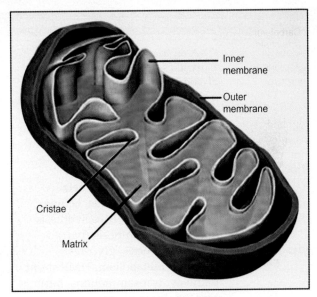

Fig. 1.3. Mitochondria

Mitochondria are spherical, oval or rod-like bodies, about 0.5 to 1 μm in diameter and up to 7 μm in length.

Erythrocytes do not contain mitochondria. The tail of spermatozoa is fully packed with mitochondria. Mitochondria are the **powerhouse of the cell**. The inner membrane contains the enzymes of **electron transport chain.** The fluid matrix contains the enzymes of citric acid cycle, urea cycle and heme synthesis.

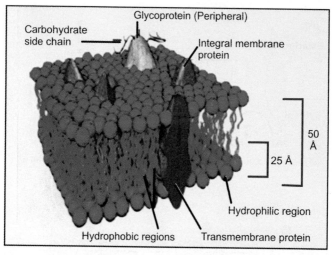

Fig. 1.4. The fluid mosaic model of membrane

The structure of the biomembranes was described as a fluid mosaic model (Singer and Nicolson, 1972). The phospholipids are arranged in bilayers with the polar head groups oriented towards the extracellular side and the cytoplasmic side with a hydrophobic core. The lipid bilayer shows free lateral movement of its components, hence the membrane is said to be **fluid in nature**.

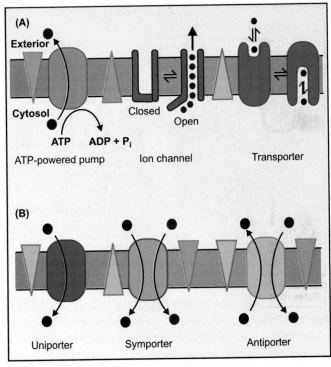

Fig. 1.5. Transport mechanisms

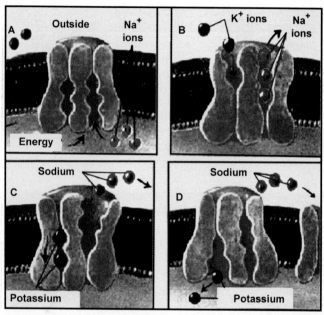

Fig. 1.6. The sodium-potassium pump. It brings sodium ions out of the cells and potassium ions into the cells. (A) Sodium ions within the cell fit into channel. (B) The channel changes shape, pumping the sodium ions. Potassium ions outside the cell move into receptor sites. (C) Sodium ions are released, potassium ions are pumped into the cell. (D) Potassium ions are released inside the cell.

Table 1.2. Comparison of cell with a factory

Plasma membrane	:	Fence with gates; gates open when message is received
Nucleus	:	Manager's office
Endo.reticulum	:	Conveyer belt of production units
Golgi apparatus	:	Packing units
Lysosomes	:	Incinerators
Vesicles	:	Lorries carrying finished products
Mitochondria	:	Power generating units

Table 1.3. Types of transport mechanisms

	Carrier	Against gradient	Energy required	Examples
Simple diffusion	No	No	Nil	Water
Facilitated diffusion	Yes	No	Nil	Glucose to RBCs
Primary active	Yes	Yes	Directly	Sodium pump
Secondary active	Yes	Yes	Indirect	Glucose to intestine
Ion channels	Yes	No	No	Sodium channel

2. Amino Acids and Proteins

Table 2.1. Classification of amino acids

A. Aliphatic amino acids
    a.   **Mono amino mono carboxylic acids:**
- Simple amino acids:
    1. Glycine
    2. Alanine
- Branched chain amino acids:
    3. Valine
    4. Leucine
    5. Isoleucine
- Hydroxy amino acids:
    6. Serine
    7. Threonine
- Sulphur containing:
    8. Cysteine
    9. Methionine
- Having amide group:
    10. Asparagine
    11. Glutamine

    b.   **Mono amino dicarboxylic acids**
        12. Aspartic acid
        13. Glutamic acid

    c.   **Di basic mono carboxylic acids**
        14. Lysine
        15. Arginine

Contd...

Contd...

B. Aromatic amino acids:
   16. Phenylalanine
   17. Tyrosine
C. Heterocyclic amino acids:
   18. Tryptophan
   19. Histidine
D. Imino acid:
   20. Proline
E. Derived amino acids:
   Hydroxy proline, hydroxy lysine, ornithine.

Table 2.2. Common amino acids

Name of amino acid	Special group present	3-letter	1-letter
Glycine		Gly	G
Alanine		Ala	A
Valine		Val	V
Leucine		Leu	L
Isoleucine		Ile	I
Serine	Hydroxyl	Ser	S
Threonine	Hydroxyl	Thr	T
Cysteine	Sulfhydryl	Cys	C
Methionine	Thioether	Met	M
Asparagine	Amide	Asn	N
Glutamine	Amide	Gln	Q

Contd...

Contd...

Name of amino acid	Special group present	3-letter	1-letter
Aspartic acid	Beta-carboxyl	Asp	D
Glutamic acid	Gamma-carboxyl	Glu	E
Lysine	Epsilon-amino	Lys	K
Arginine	Guanidinium	Arg	R
Phenylalanine	Benzene	Phe	F
Tyrosine	Phenol	Tyr	Y
Tryptophan	Indole	Trp	W
Histidine	Imidazole	His	H
Proline (imino acid)	Pyrrolidine	Pro	P

Table 2.3. Essential amino acids

Arginine, **H**istidine, **I**soleucine, **L**eucine, **T**hreonine, **L**ysine, **M**ethionine, **P**henylalanine, **T**ryptophan and **V**aline.

Arginine and histidine are semi-essential amino acids; while others are essential.

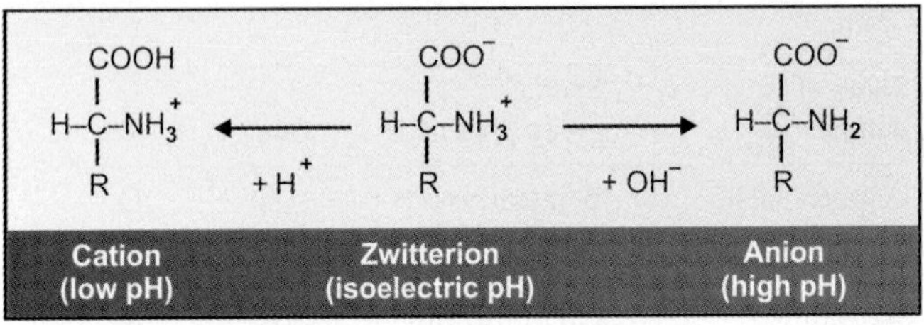

Fig. 2.1. Ionic forms of amino acids

Iso-electric Point

Amino acids can exist as **ampholytes** or **zwitterions** in solution, depending on the pH of the medium. The pH at which the molecule carries no net charge is known as *iso-electric point or iso-electric pH* (pI). In acidic solution they are cationic in form and in alkaline solution they behave as anions.

At iso-electric point the amino acid will **carry no net charge;** all the groups are ionized but the charges will cancel each other. At iso-electric point, there is *no mobility in an electrical field*. Solubility and buffering capacity will be minimum at iso-electric pH.

At a particular pH, 50% of the molecules are in cation form and 50% in zwitterion form. This pH is pK_1 (with regard to COOH). When 50% of molecules are anions, that pH is called pK_2 (with respect to NH_2). The buffering action is maximum in and around pK_1 or at pK_2 and minimum at pI.

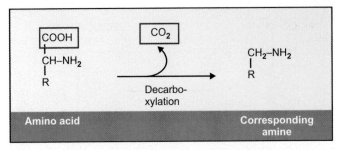

Fig. 2.2. Decarboxylation of amino acid

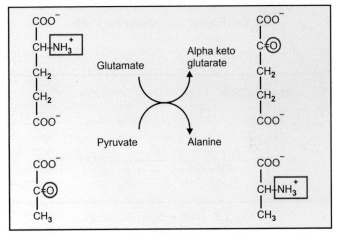

Fig. 2.3. Transamination reaction

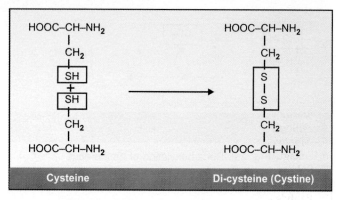

Fig. 2.4. Formation of disulphide bridges

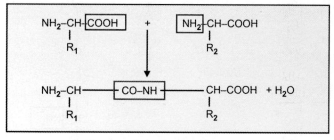

Fig. 2.5. Peptide bond formation

Table 2.4. Color reactions of amino acids

Reactions	Answered by specific group
1. Ninhydrin	Alpha amino group
2. Biuret reaction	Peptide bonds
3. Xanthoproteic test	Benzene ring (Phe, Tyr, Trp)
4. Millon's test	Phenol (Tyrosine)
5. Aldehyde test	Indole (Tryptophan)
6. Sakaguchi's test	Guanidium (Arginine)
7. Sulphur test	Sulfhydryl (Cysteine)
8. Nitroprusside test	Sulfhydryl (Cysteine)
9. Pauly's test	Imidazole (Histidine)

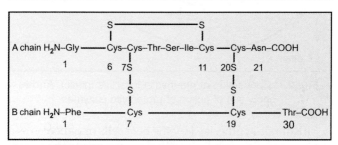

Fig. 2.6. Primary structure of human insulin

Insulin has **two polypeptide chains;** the A chain (Glycine chain) with 21 amino acids and B (Phenyl alanine) chain with 30 amino acids. They are held together by a pair of **disulphide bonds**.

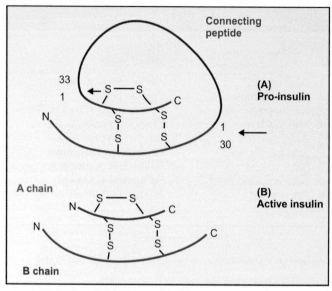

Fig. 2.7. Conversion of pro-insulin to active insulin. Arrows show site of action of proteolytic enzymes

Insulin is synthesised by the beta cells of pancreas as a prohormone; proinsulin is a **single polypeptide chain** with 86 amino acids. Biologically active insulin with 2 chains is formed by removal of the central portion of the proinsulin before release. This cleavage occurs at specific sites and the peptide thus removed is called the **C-peptide** (connecting peptide).

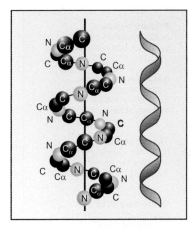

Fig. 2.8. Alpha helix of proteins. Vertical line is the axis of helix. Each turn is formed by 3.6 amino acid residues. The distance between each amino acid (translation) is 1.5 Å

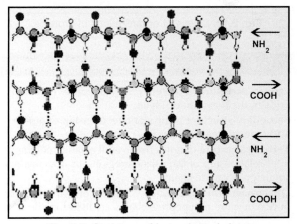

Fig. 2.9. Structure of beta-pleated sheet

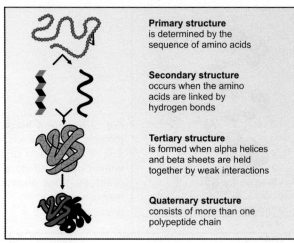

Primary structure
is determined by the
sequence of amino acids

Secondary structure
occurs when the amino
acids are linked by
hydrogen bonds

Tertiary structure
is formed when alpha helices
and beta sheets are held
together by weak interactions

Quaternary structure
consists of more than one
polypeptide chain

Fig. 2.10. Levels of organizations of proteins

Table 2.5. Definitions of levels of organization

1. **Primary structure** of protein means the order of amino acids in the polypeptide chain.
2. **Secondary structure** is the steric relationship of amino acids, close to each other.
3. **Tertiary structure** denotes the overall arrangement and inter-relationship of the various regions, or domains of a single polypeptide chain.
4. **Quaternary structure** results when the proteins consisting of two or more polypeptide chains are held together by non-covalent forces.

Table 2.6. Classification of proteins

Simple Proteins
  i. Albumins: They are soluble in water and coagulated by heat.
 ii. Globulins: These are insoluble in pure water, but soluble in dilute salt solutions.
iii. Protamines: These are soluble in water, dilute acids and alkalies.
 iv. Prolamines: They are soluble in 70 to 80% alcohol, but insoluble in pure water.
  v. Lectins: They are proteins having high affinity to sugar groups. Lectins are usually precipitated by 30 to 60% ammonium sulphate.
 vi. Scleroproteins: They are insoluble in water, salt solutions and organic solvents and soluble only in hot strong acids.

Conjugated Proteins
They are combinations of protein with a non-protein part, called prosthetic group.
  i. Glycoproteins
 ii. Lipoproteins
iii. Nucleoproteins
 iv. Chromoproteins
  v. Phosphoproteins
 vi. Metalloproteins

Derived Proteins
They are degradation products of native proteins.

3. Enzymes

Table 3.1. Classification of enzymes

Class 1. Oxidoreductases: Transfer of hydrogen; e.g. alcohol dehydrogenase.

Class 2. Transferases: Transfer of groups other than hydrogen (Subclass: Kinase, transfer of phosphoryl group from ATP; e.g., hexokinase)

Class 3. Hydrolases: Cleave bond; add water; e.g., acetyl choline esterase

Class 4. Lyases: Cleave without adding water, e.g., aldolase. (Subclass: Hydratase; add water to double bond)

Class 5. Isomerases: Intramolecular transfers. Example, triose phosphate isomerase.

Class 6. Ligases: ATP dependent condensation of two molecules, e.g., acetyl CoA carboxylase

Table 3.2. Examples of co-enzymes

Co-enzyme	Group transferred
Thiamine pyrophosphate (TPP)	Hydroxy ethyl
Pyridoxal phosphate (PLP)	Amino group
Biotin	Carbon dioxide
Coenzyme-A (Co-A)	Acyl groups
Tetrahydrofolate (FH4)	One carbon groups
Adenosine triphosphate (ATP)	Phosphate

Table 3.3. Metallo-enzymes

Metal	Enzyme containing the metal
Zinc	Carbonic anhydrase, carboxy peptidase, alkaline phosphatase
Magnesium	Hexokinase, phosphofructokinase, enolase
Manganese	Hexokinase, enolase
Copper	Tyrosinase, cytochrome oxidase,
Iron	Cytochrome oxidase, xanthine oxidase
Calcium	Lecithinase, lipase
Molybdenum	Xanthine oxidase

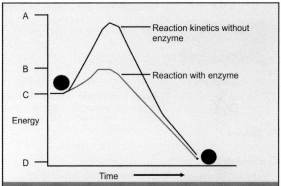

Red circle = substrate; black circle = product; C = energy level of substrate; D = energy level of product; C to A = activation energy in the absence of enzyme; C to B is activation energy in presence of enzyme; B to A = lowering of activation energy by enzyme

Fig. 3.1. Lowering of activation energy by enzymes

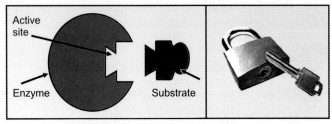

Fig. 3.2. Enzyme and substrate are specific to each other. This is similar to key and lock (Fischer's theory)

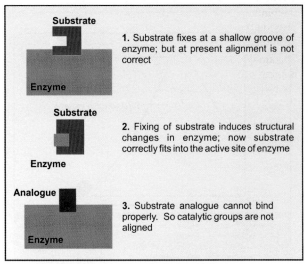

Fig. 3.3. Koshland's induced fit theory

Table 3.4. Salient features of active site or active centre of enzyme

1. That area of the enzyme where catalysis occurs is referred to as active site or active centre.

2. Although all parts are required for keeping the exact three- dimensional structure of the enzyme, the reaction is taking place at the active site.

3. Generally active site is situated in a crevice or cleft of the enzyme molecule.

4. To the active site, the specific substrate is bound. During the binding, these groups may realign themselves to provide the unique conformational orientation so as to promote exact fitting of substrate to the active site.

5. The amino acids or groups that directly participate in making or breaking the bonds (present at the active site) are called catalytic residues or catalytic groups.

6. Proteolytic enzymes having a serine residue at the active centre are called serine proteases, e.g. trypsin, chymotrypsin and coagulation factors.

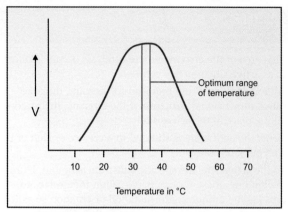

Fig. 3.4. Effect of temperature on velocity

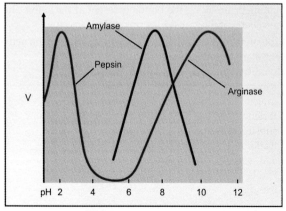

Fig. 3.5. Effect of pH on enzyme velocity

Table 3.5. Derivation of equilibrium constant

$$V \; \alpha \; [A] \, [B]$$

At equilibrium, forward reaction and backward reaction are equal, so that

$$A + B \underset{K_2}{\overset{K_1}{\rightleftharpoons}} C + D$$

Forward reaction R_1 $= K_1 \, [A] \, [B]$
and backward reaction R_2 $= K_2 \, [C] \, [D]$
At equilibrium, R_1 $= R_2$
Or, $K_1 \, [A] \, [B]$ $= K_2 \, [C] \, [D]$

Or, $\dfrac{K_1}{K_2} = \dfrac{[C] \, [D]}{[A] \, [B]}$ = Keq or Equilibrium constant.

Table 3.6. Michaelis costant (Km)

 i. Km value is substrate concentration (expressed in moles/L) at half-maximal velocity.
 ii. It denotes that 50% of enzyme molecules are bound with substrate molecules at that particular substrate concentration.
iii. Km is independent of enzyme concentration.
 iv. Km is the signature of the enzyme.
 v. Km denotes the affinity of enzyme to substrate.

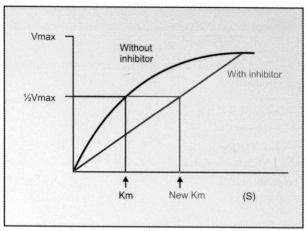

Fig. 3.6. Substrate saturation curve in presence and absence of competitive inhibitor

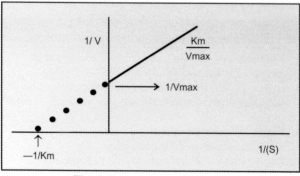

Fig. 3.7. Lineweaver-Burk plot

Table 3.7. Comparison of two types of inhibition

	Competitive inhibition	Non-competitive inhibition
Structure of inhibitor	Substrate analogue	Unrelated molecule
Inhibition	Reversible	Generally irreversible
Excess substrate	Inhibition relieved	No effect
Km	Increased	No change
Vmax	No change	Decreased
Significance	Drug action	Toxicological

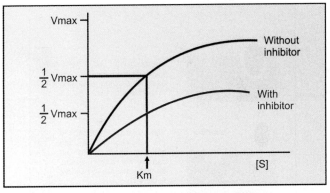

Fig. 3.8. Non-competitive inhibition

Table 3.8. Examples of allosteric enzymes

Enzyme	Allosteric inhibitor	Allosteric activator
1. ALA synthase	Heme	
2. Aspartate transcarbamoylase	CTP	ATP
3. HMGCoA-reductase	Cholesterol	
4. Phospho-fructokinase	ATP, citrate	AMP, F-2, 6-P

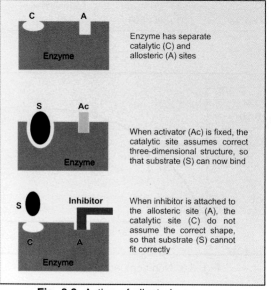

Fig. 3.9. Action of allosteric enzymes

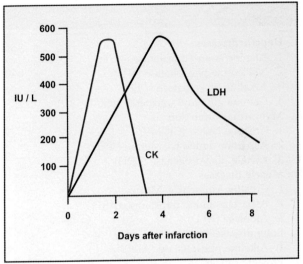

Fig. 3.10. Time course of elevation of lactate dehydrogenase (LDH) and creatine kinase (CK) in myocardial infarction

In myocardial infarction, CK is the first enzyme increased in myocardial infarction. CK-MB is the isoenzyme from cardiac muscle. Total LDH activity is increased, while H_4 iso-enzyme is increased 5 to 10 times more. The magnitude of the peak value as well as the area under the graph will be roughly proportional to the size of the myocardial infarct.

Table 3.9. Enzyme profiles in disease

I. **Hepatic diseases**
   1. Alanine amino transferase (ALT)
   2. Nucleotide phosphatase (NTP)
   3. Alkaline phsophatase (ALP)
   4. Gamma glutamyl transferase (GGT)
II. **Myocardial infarction**
   1. Creatine kinase (CK-MB)
   2. Aspartate amino transferase (AST)
   3. Lactate dehydrogenase (LDH)
III. **Muscle diseases**
   1. Creatine kinase (CK-MM)
   2. Aspartate amino transferase (AST)
   3. Aldolase (ALD)
IV. **Bone diseases**
   1. Alkaline phosphatase (ALP)
V. **Prostate cancer**
   1. Prostate specific antigen (PSA)
   2. Acid phosphatase (ACP)

Table 3.10. Therapeutic use of enzymes

1. Asparaginase	Acute lymphoblastic leukemia
2. Streptokinase	To lyse intravascular clot
3. Hyaluronidase	Enhances local anesthetics
4. Pancreatin (trypsin and lipase)	Pancreatic insufficiency; oral administration
5. Alpha 1-antitrypsin	AAT deficiency; emphysema

4. Carbohydrate Chemistry

Table 4.1. Common monosaccharides

No. of carbon atoms	Generic name	Aldoses (with aldehyde group)	Ketoses (with keto group)
3	Triose	Ex: Glyceraldehyde	Ex: Dihydroxyacetone
4	Tetrose	Erythrose	Erythrulose
5	Pentose	Arabinose Xylose Ribose	Xylulose Ribulose
6	Hexose	Glucose Galactose Mannose	Fructose

Table 4.2. Hexoses of physiological importance

Glucose	Blood sugar. Main source of energy in body
Fructose	Constituent of sucrose, the common sugar
Galactose	Constituent of lactose, glycolipids and glycoproteins
Mannose	Constituent of globulins, mucoproteins and glycoproteins

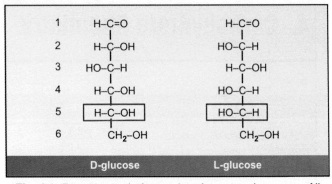

Fig. 4.1. Penultimate (reference) carbon atom in sugars. All naturally occurring sugars are D sugars

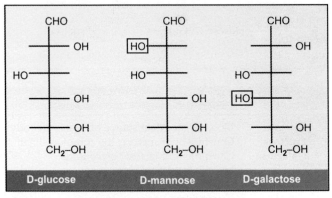

Fig. 4.2. Epimers of D-glucose

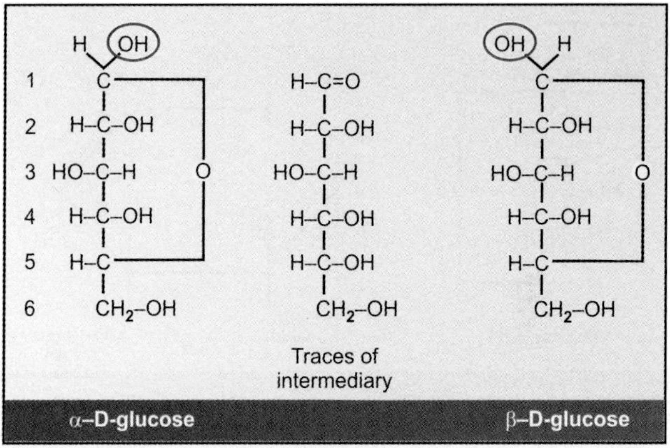

Fig. 4.3. Anomers of D-glucose

When D glucose is crystallised at room temperature, and a fresh solution is prepared, its specific rotation of polarised light is +112°; but after 12 to 18 hours it changes to +52.5°. This change in rotation with time is called **mutarotation.** This is explained by the fact that D-glucose has two *anomers,* **alpha and beta varieties.** alpha-D-glucose has specific rotation of +112° and beta-D-glucose has +19°. Both undergo mutarotation and at equilibrium one-third molecules are alpha type and 2/3rd are beta variety to get the specific rotation of +52.5°. These anomers are produced by the spatial configuration with *reference to the first carbon atom in aldoses and second carbon atom in ketoses.* Hence, these carbon atoms are known as **anomeric carbon atoms**.

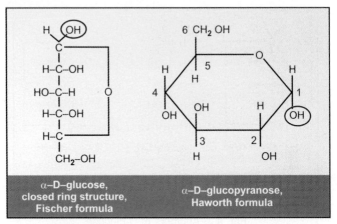

α–D–glucose, closed ring structure, Fischer formula

α–D–glucopyranose, Haworth formula

Fig. 4.4. Different representations of D-glucose

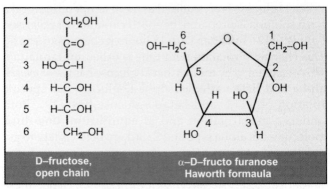

D–fructose, open chain

α–D–fructo furanose Haworth formaula

Fig. 4.5. Different representations of D-fructose

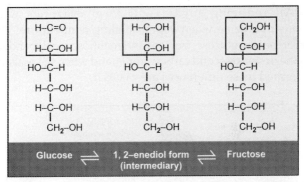

Fig. 4.6. Interconversion of sugars

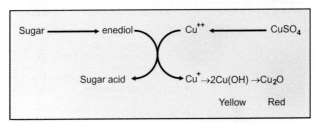

Fig. 4.7. Benedict's test, principle

Benedict's reagent is very commonly employed to detect the presence of glucose in urine (glucosuria). Benedict's reagent contains sodium carbonate, copper sulphate and sodium citrate. In the alkaline medium provided by sodium carbonate, the copper remains as cupric hydroxide. Sodium citrate acts as a stabilising agent to prevent precipitation of cupric hydroxide.

All reducing sugars will form osazones with phenylhydrazine when kept at boiling temperature. The differences in glucose, fructose and mannose are dependent on the first and second carbon atoms, and when the osazone is formed these differences are masked.

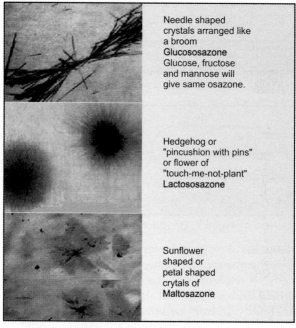

Needle shaped crystals arranged like a broom **Glucososazone** Glucose, fructose and mannose will give same osazone.

Hedgehog or "pincushion with pins" or flower of "touch-me-not-plant" **Lactososazone**

Sunflower shaped or petal shaped crytals of **Maltosazone**

Fig. 4.8. Shape of osazones under microscope

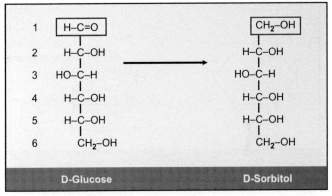

Fig. 4.9. Reduction of sugar to alcohol

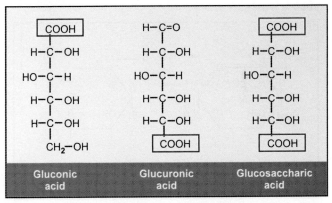

Fig. 4.10. Oxidative products of glucose

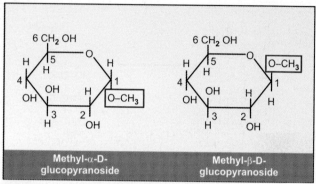

Fig. 4.11. Glycosides. When the hemi-acetal group is condensed with an alcohol or phenol group, it is called a glycoside. Glycosides do not reduce Benedict's reagent, because the sugar group is masked

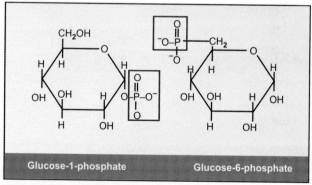

Fig. 4.12. Phosphorylated sugars. Metabolism of sugars inside the body starts with phosphorylation

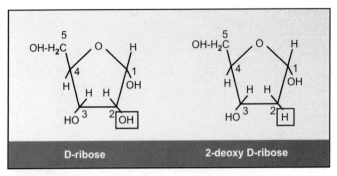

Fig. 4.13. Sugars of nucleic acids

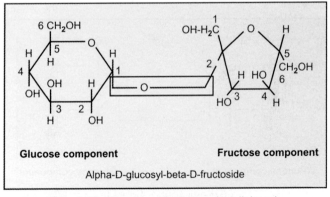

Alpha-D-glucosyl-beta-D-fructoside

Fig. 4.14. Structure of sucrose (1-2 linkage)

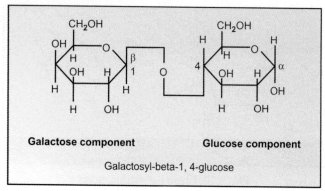

Galactose component   **Glucose component**

Galactosyl-beta-1, 4-glucose

Fig. 4.15. Lactose

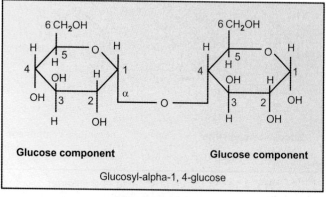

Glucose component   **Glucose component**

Glucosyl-alpha-1, 4-glucose

Fig. 4.16. Maltose

Table 4.3. Salient features of important sugars

Monosaccharides
Glucose	Aldohexose
Galactose	4th epimer of glucose
Mannose	2nd epimer of glucose
Fructose	Ketohexose

Disaccharides
Glucose + Galactose	=	Lactose (reducing)
Glucose + Glucose	=	Maltose (reducing)
Glucose + Fructose	=	Sucrose (nonreducing)

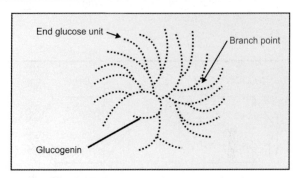

Fig. 4.17. Branched glycogen molecule

Glycogen is the reserve carbohydrate in animals. It is stored in liver and muscle. Glycogen is composed of glucose units joined by alpha-1,4 and alpha-1,6 glycosidic linkages. Glycogen is branched. The branching points are made by alpha-1, 6 linkages.

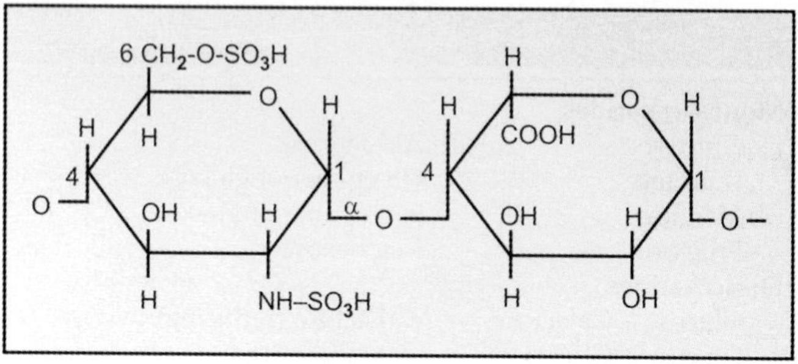

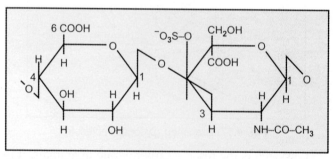

Fig. 4.18. Sulphated glucosamine-alpha-1, 4-iduronic acid. Repeating units in heparin

Fig. 4.19. D-glucuronic acid-beta-1, 3-N-acetyl galactosamine-4 sulfate. Repeating units of chondroitin sulphate

Table 4.4. Mucopolysaccharides

Mucopolysaccharides or glycosaminoglycans (GAG) are carbohydrates containing uronic acid and amino sugars. They are present in connective tissues, tendons, synovial fluid and vitreous humor. Some examples of mucopolysaccharides are given below.

1. Hyaluronic acid
It is composed of repeating units of N-acetyl-glucosamine → beta -1,4-glucuronic acid → beta-1-3-N-acetyl glucosamine and so on.

2. Heparin
It contains repeating units of sulphated glucosamine → alpha-1, 4-L-iduronic acid → and so on. (Fig. 4.18).

3. Chondroitin sulphate
It is composed of repeating units of glucuronic acid → beta-1, 3-N-acetyl galactosamine sulphate → beta-1, 4 and so on (Fig. 4.19).

4. Keratan sulphate
It is the only GAG which **does not contain any uronic acid.** The repeating units are galactose and N-acetyl glucosamine in beta linkage.

5. Chemistry of Lipids

Table 5.1. Classification of lipids

I. Simple lipids
   a. Triacyl glycerol or triglycerides or neutral fat
   b. Waxes
II. Compound lipids
A. Phospholipids, containing phosphoric acid.
   1. Nitrogen containing glycerophosphatides:
      i. Lecithin (phosphatidyl choline)
      ii. Cephalin (phosphatidyl ethanol amine)
      iii. Phosphatidyl serine
   2. Non-nitrogen glycerophosphatides
      i. Phosphatidyl inositol
      ii. Phosphatidyl glycerol
      iii. Diphosphatidyl glycerol (cardiolipin)
   3. Plasmalogens, having long chain alcohol
      i. Choline plasmalogen
      ii. Ethanolamine plasmalogen
   4. Phosphosphingosides, with sphingosine
         Sphingomyelin

Contd...

Contd...

B. Non-phosphorylated lipids
  1. Glycosphingolipids (carbohydrate)
     i. Cerebrosides (ceramide monohexosides)
     ii. Globosides (ceramide oligosaccharides)
     iii. Gangliosides
  2. Sulpholipids or sulfatides
     i. Sulphated cerebrosides
     ii. Sulphated globosides
     iii. Sulphated gangliosides
III. Derived lipids
   Fatty acids, steroids, prostaglandins, leukotrienes,
   terpenes, dolichols, etc.
IV. Lipids complexed to other compounds
   Proteolipids and lipoproteins.

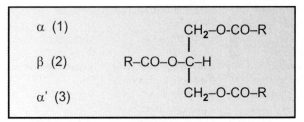

Fig. 5.1. Triacyl glycerol (TAG) (Triglyceride)

Table 5.2. Characteristics of common fatty acids

Common name	No carbon atoms	Chemical nature	Occurrence
A. Even chain, saturated fatty acids			
Acetic	2	Saturated; small chain	Vinegar
Butyric	4	Do	Butter
Caproic	6	Do	Butter
Capric	10	Do	Coconut oil
Lauric	12	Do	Coconut oil
Myristic	14	Do	Coconut oil
Palmitic	**16**	**Saturated; long chain**	**Body fat**
Stearic	**18**	**Do**	**Do**
Arachidic	20	Do	Peanut oil (Arachis oil)
B. Odd-chain fatty acids			
Propionic	3	Saturated; Odd chain	Metabolism
C. Even chain, unsaturated fatty acids			
Palmitoleic	16	Monounsaturated (ω 7)	Body fat
Oleic	18	Do (ω 9)	Do
Erucic	22	Do (ω 9)	Mustard oil
Nervonic	24	Do (ω 9)	Brain lipids
Linoleic	**18**	**2 double bonds (ω 6)**	**Vegetable oils**
Linolenic	**18**	**3 double bonds (ω 3)**	**Do**
Arachidonic	**20**	**4 double bonds (ω 6)**	**Vegetable oils**
Timnodonic	20	Eicosa pentaenoic (ω 3)	Fish oils, brain
Clupanodonic	22	Docosa pentaenoic (ω 3)	Fish oils, brain
Cervonic	22	Docosa hexaenoic (ω 3)	Fish oils, brain
D. Branched fatty acids			
Iso valeric acid	5	Branched	Metabolic intermediate

Table 5.3. Composition of oils and fats

Name	Saturated fatty acids(%)	Mono-unsaturated fatty acids(%)	PUFA (%)
Coconut oil	(*) 86	12	2
Groundnut oil	18	46	36
Gingelly oil (Til oil)	13	50	37
Palm oil	42	52	6
Cotton seed oil	26	19	55
Mustard oil (rapeseed)	34	48	18
Safflower oil (Kardi)	9	12	79
Sunflower oil	12	24	64
Butter	75	20	5
Ox (Tallow)	53	42	5
Pig (Lard)	42	46	12
Fish oil	30	13	57

(*) these saturated fatty acids are medium chain fatty acids.

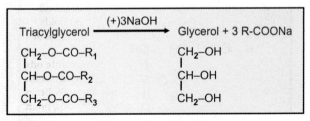

Fig. 5.2. Saponification

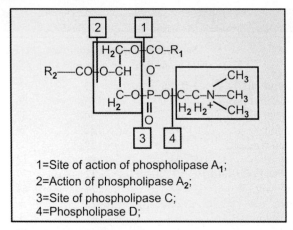

1=Site of action of phospholipase A$_1$;
2=Action of phospholipase A$_2$;
3=Site of phospholipase C;
4=Phospholipase D;

Fig. 5.3. Lecithin. R$_1$ and R$_2$ are fatty acids. Blue square depicts glycerol group. The red square is choline which shows polar or hydrophilic property

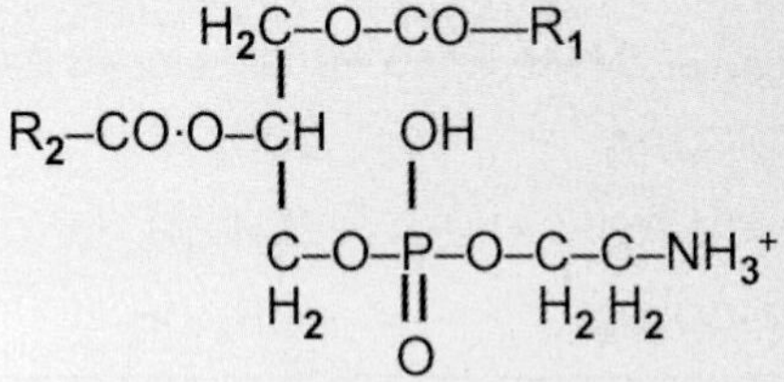

Fig. 5.4. Cephalin (Phosphatidyl ethanolamine)

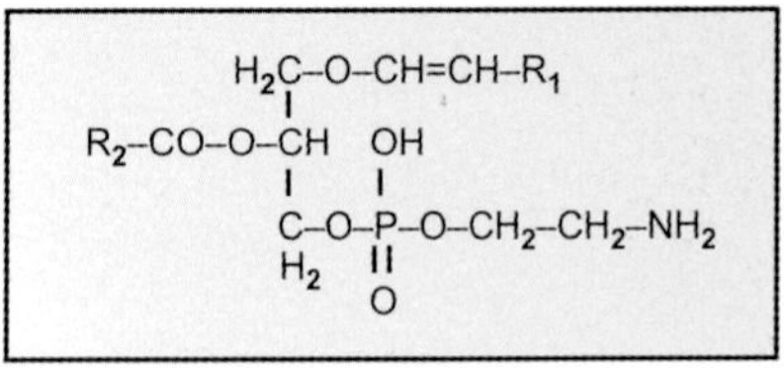

$$H_2C-O-CH=CH-R_1$$
$$R_2-CO-O-CH \quad OH$$
$$\underset{H_2}{C}-\overset{\overset{O}{\parallel}}{P}-O-CH_2-CH_2-NH_2$$
$$O$$

Fig. 5.5. Ethanolamine plasmalogen

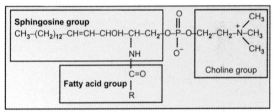

Sphingosine group
$$CH_3-(CH_2)_{12}-CH=CH-CHOH-CH-CH_2-O-\overset{\overset{O}{\parallel}}{\underset{O^-}{P}}-O-CH_2-CH_2-\overset{+}{N}-CH_3$$

NH

Choline group

Fatty acid group
C=O
R

Fig. 5.6. Sphingomyelin

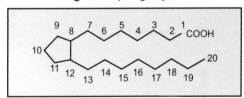

Fig. 5.7. Prostanoic acid

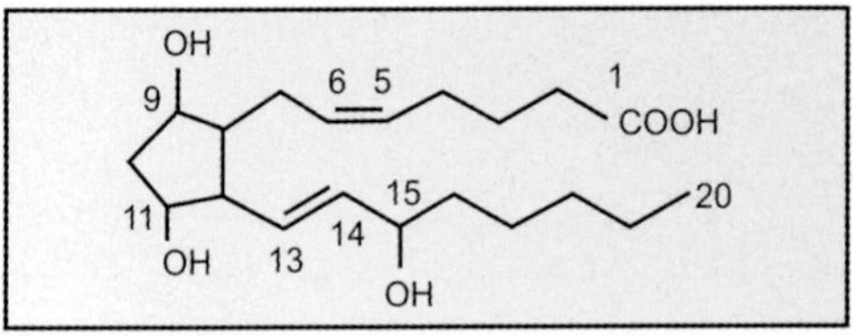

Fig. 5.8. Prostaglandin F2

Section 2

Carbohydrate Metabolism

6. Glycolysis

Table 6.1. Glucose transporters

Transporters	Present in	Properties
GluT1	RBC, brain, kidney	Glucose uptake in most of cells
GluT2	Serosal surface of intestinal cells, beta cell pancreas	Glucose uptake in liver; Glucose sensor in beta cells
GluT3	Neurons, brain	Glucose into brain
GluT4	Skeletal, heart muscle, adipose tissue	Insulin mediated glucose uptake

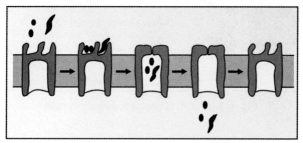

Fig. 6.1. SGluT. Sodium and glucose co-transport system at luminal side; sodium is then pumped out

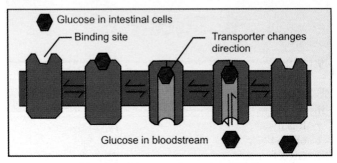

Fig. 6.2. Glucose absorption (Glu2)

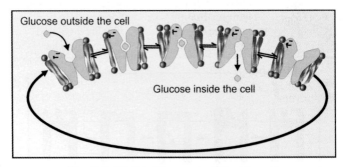

Fig. 6.3. GluT4. Glucose transport in cells

Table 6.2. Importance of glycolysis or (Embden-Meyerhof pathway)

  i. In this pathway glucose is converted to **pyruvate** (aerobic condition) or **lactate** (anaerobic condition), along with production of a small quantity of energy.

  ii. All the reaction steps take place in the cytoplasm. It is the only pathway that is taking place in all the cells of the body.

 iii. Glycolysis is the only source of energy in erythrocytes.

  iv. In strenuous exercise, when muscle tissue lacks enough oxygen, *anaerobic glycolysis forms the major source of energy for muscles.*

  v. The glycolytic pathway may be considered as the preliminary step before complete oxidation.

  vi. The glycolytic pathway also provides carbon skeletons for synthesis of certain non-essential amino acids as well as glycerol part of fat.

 vii. Most of the reactions of the glycolytic pathway are reversible, which are also used for gluconeogenesis.

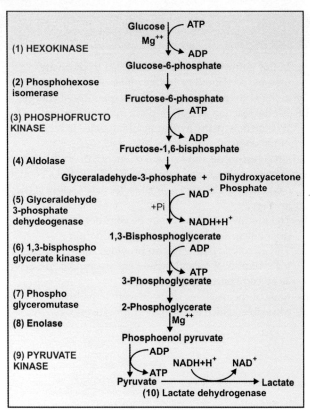

Fig. 6.4. Glycolysis (Embden-Meyerhof) pathway. Steps 1, 3 and 9 are key enzyems; these are irreversible. Steps 5, 6 and 9 produce energy

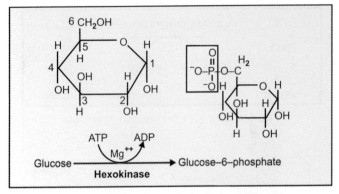

Fig. 6.5. Step 1 of glycolysis; irreversible step

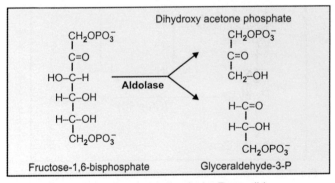

Fig. 6.6. Step 4 of glycolysis; Reversible

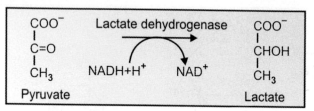

Fig. 6.7. Step 10; LDH reaction; reversible

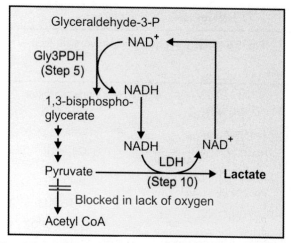

Fig. 6.8. Lactate formation is necessary for reconversion of NADH to NAD$^+$ during anaerobiasis

Table 6.3. Energy yield (number of ATP generated) per molecule of glucose in the glycolytic pathway, under anaerobic conditions (Oxygen deficiency)

Step	Enzyme	Source	No. of ATPs gained per glucose mol
1	Hexokinase	–	Minus 1
3	Phosphofructokinase	–	Minus 1
6	1,3-bisphosphogly-cerate kinase	ATP	$1 \times 2 = 2$
9	Pyruvate kinase	ATP	$1 \times 2 = 2$
			Total = 4 minus 2 = 2

Table 6.4. Energy yield (number of ATP generated) per molecule of glucose in the glycolytic pathway, under *aerobic conditions* (oxygen is available)

Step	Enzyme	Source	No. of ATPs gained per glucose mol
1	Hexokinase	–	Minus 1
3	Phosphofructokinase	–	Minus 1
5	Glyceraldehyde-3-phosphate dehydrogenase	NADH	$3 \times 2 = 6$
6	1,3-bisphosphogly-cerate kinase	ATP	$1 \times 2 = 2$
9	Pyruvate kinase	ATP	$1 \times 2 = 2$
		Total = 10 minus 2	= 8

Table 6.5. Energy yield (number of ATP generated) per molecule of glucose when it completely oxidised through glycolysis plus citric acid cycle, under *aerobic conditions*.

Pathway		Step enzyme	Source	No of ATPs gained per glucose
Glycolysis	1	Hexokinase	–	Minus 1
Do	3	Phospho-fructokinase	–	Minus 1
Do	5	Glyceralde-hyde-3-P DH	NADH	$3 \times 2 = 6$
Do	6	1,3-BPG kinase	ATP	$1 \times 2 = 2$
Do	9	Pyruvate kinase	ATP	$1 \times 2 = 2$
Pyruvate to acetyl CoA		Pyruvate dehydrogenase	NADH	$3 \times 2 = 6$
TCA cycle	3	Isocitrate DH	NADH	$3 \times 2 = 6$
Do	4	Alpha ketoglutarate dehydrogenase	NADH	$3 \times 2 = 6$
Do	5	Succinate thiokinase	GTP	$1 \times 2 = 2$
Do	6	Succinate DH	$FADH_2$	$2 \times 2 = 4$
Do	8	Malate dehydrogenase	NADH	$3 \times 2 = 6$

Net generation in glycolytic pathway 10 minus 2	= 8
Generation in pyruvate dehydrogenation	= 6
Generation in citric acid cycle	= 24
Net generation of ATP from one glucose mol	= 38

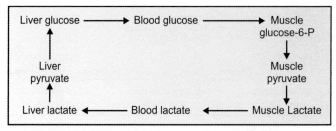

Fig. 6.9. Cori's cycle. Contracting muscle has lack of oxygen. So pyruvate is made to lactate. This can be reconverted to glucose in liver having plenty of oxygen

Table 6.6. Regulatory enzymes of glycolysis		
Enzyme	*Activation*	*Inhibition*
HK		G-6-P
GK	Insulin	Glucagon
PFK	Insulin, AMP	Glucagon, ATP
	F-6-P, PFK-2	Citrate, low pH
	F2,6-BP	CyclicAMP
PK	Insulin, F1,6-BP	Glucagon, ATP
		CyclicAMP
PDH	CoA, NAD	Acetyl CoA, NADH

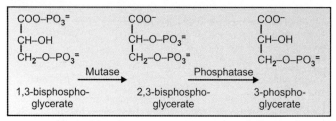

Fig. 6.10. BPG shunt; step 6 of glycolysis is bypassed in erythrocytes

The 2,3-BPG when combines with hemoglobin, **reduces the affinity towards oxygen**. *Under hypoxic conditions the 2,3-BPG concentration in the RBC increases,* thus favouring the release of oxygen to the tissues even when pO_2 is low. The compensatory increase in 2,3-BPG in high altitudes and fetal tissues also favours oxygen dissociation.

Table 6.7. Key enzymes	
Irreversible steps in glycolysis	*Corresponding key gluconeogenic enzymes*
Pyruvate kinase (step 9)	Pyruvate carboxylase; Phosphoenol pyruvate-carboxy kinase
Phosphofructokinase (step 3)	Fructose-1,6-bisphosphatase
Hexokinase (step 1)	Glucose-6-phosphatase

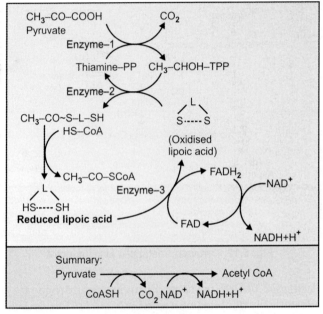

Fig. 6.11. Pyruvate dehydrogenase reaction. Completely irreversible process

Enzyme 1 = Pyruvate dehydrogenase

Enzyme 2 = Dihydrolipoyl transacetylase

Enzyme 3 = Dihydrolipoyl dehydrogenase

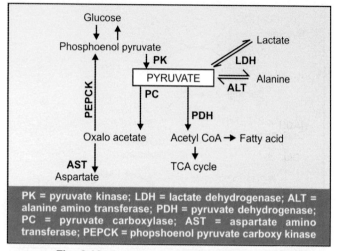

PK = pyruvate kinase; LDH = lactate dehydrogenase; ALT = alanine amino transferase; PDH = pyruvate dehydrogenase; PC = pyruvate carboxylase; AST = aspartate amino transferase; PEPCK = phopshoenol pyruvate carboxy kinase

Fig. 6.12. Pyruvate; metabolic junction point

Regulation of PDH reaction: Pyruvate dehydrogenase is allosterically inhibited by acetyl CoA and NADH. It is also inhibited by ATP. Insulin activates PDH, especially in adipose tissue.

PDH is a completely irreversible process: The oxidative decarboxylation of pyruvate to acetyl CoA is a completely irreversible process. There are no pathways available in the body to circumvent this step.

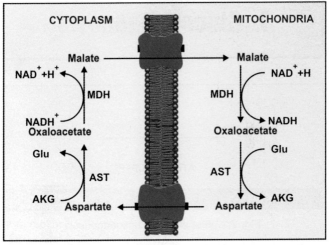

Fig. 6.13. Malate-aspartate shuttle. MDH = malate dehydro-genase. AST = aspartate amino transferase. Glu= glutamic acid. AKG = alpha ketoglutaric acid

In the first reaction of gluconeogenes, carboxylation of pyruvate to oxaloacetate is catalysed by a mitochondrial enzyme, pyruvate carboxylase (See Fig. 7.1). It contains **biotin** which acts as a carrier of active CO_2. The reaction requires ATP. The rest of the reactions of gluconeogenesis are taking place in cytosol. Hence the oxaloacetate has to be transported from mitochondria to cytosol. This is achieved by the *malate shuttle* (see Fig. 6.13).

7. Gluconeogenesis

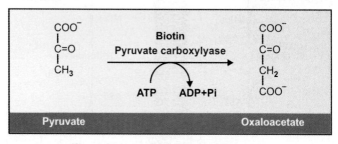

Fig. 7.1. First step in gluconeogenesis

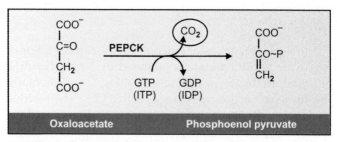

Fig. 7.2. PEPCK = Phosphoenol pyruvate carboxy kinase

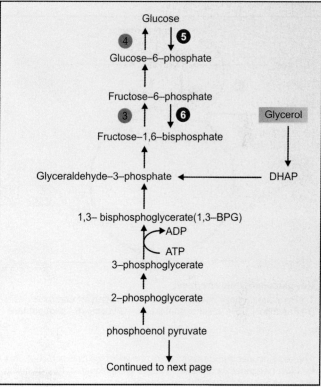

Fig. 7.3.

Contd...

Contd...

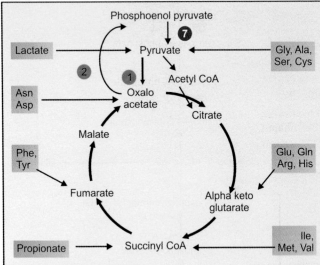

Key gluconeogenic enzymes:

1 = Pyruvate carboxylase; 2 = Phospho enol pyruvate carboxykinase;
3 = Fructose – 1,6 – bisphosphatase; 4 = Glucose – 6 – phosphatase

Key glycolytic enzymes:

5 = Hexokinase; 6 = Phosphofructokinase; 7 = Pyruvate kinase.
The key gluconeogenic enzymes are shown in blue circles. Substrates
for gluconeogensesis are shown inside green squares. Key glycolitic
enzymes (irreversible steps) are in red circles.

Fig. 7.3. Gluconeogenic pathway

Table 7.1. Regulatory enzymes of gluconeogenesis (compare with Table 6.6)

Enzyme	Activation	Inhibition
PC	Cortisol, Glucagon	Insulin, ADP
PEPCK	Do	Insulin
Fructose-1,6-bis-phosphatase	Do	F-1,6-BP, AMP
G-6-phosphatase	Do	Insulin

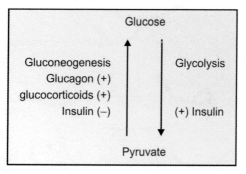

Fig. 7.4. Hormonal regulation of gluconeogenesis

8. Glycogenolysis

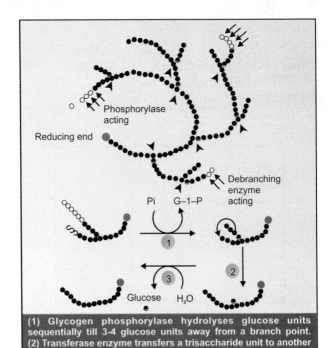

(1) Glycogen phosphorylase hydrolyses glucose units sequentially till 3-4 glucose units away from a branch point. (2) Transferase enzyme transfers a trisaccharide unit to another branch. (3) Remaining one unit is removed by $\alpha-1-6$ glucosidase.

Fig. 8.1. Glycogenolysis

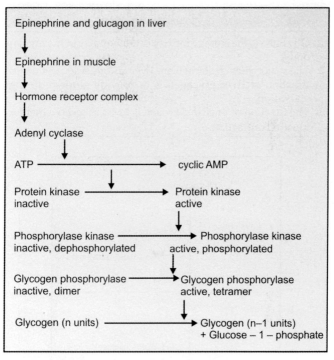

Fig. 8.2. Cyclic AMP mediated degradation of glycogen

Table 8.1. Functions of glycogen

1. Glycogen is the storage form of carbohydrates in the human body.
2. The major sites of storage are liver and muscle. The major function of **liver glycogen** is to provide glucose during starvation.
3. The function of muscle glycogen is to act as reserve fuel for muscle contraction.

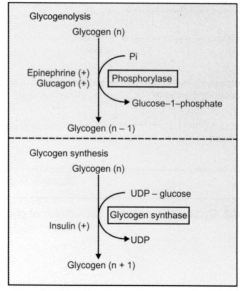

Fig. 8.3. Effects of hormones on glycogen

Table 8.2. Glycogen storage diseases

Disease	Deficient enzyme	Salient features
Type I Von Gierke's disease	Glucose-6-phosphatase	Fasting hypoglycemia; Hepatomegaly
Type II Generalised glycogenosis; Pompe's disease	Lysosomal maltase	Liver, heart and muscle affected; death before 2 years
Type III Limit dextrinosis Cori's disease	Debranching enzyme	Highly branched dextrin accumulates; Hepatomegaly
Type IV Amylopectinosis Anderson's disease	Branching enzyme	Glycogen with little branches; hepatomegaly
Type V McArdle's disease	Muscle phosphorylase	Exercise intolerance
Type VI Hers disease	Liver phosphorylase	Hypoglycemia
Type VII Tarui's disease	Muscle PFK	Accumulation of glycogen in muscles
Type VIII	Liver phosphorylase kinase	
Type IX Lewis disease	Glycogen synthase	

9. Minor Pathways of Carbohydrates

Table 9.1. Significance of the pathway

Hexose monophosphate (HMP) pathway is also known as: Pentose phsophate pathway; Dickens-Horecker pathway; Shunt pathway; or Phospho-gluconate oxidative pathway

1. **Tissues**

   It is seen in organs where fatty acid or steroid synthesis is taking place, such as in liver, mammary glands, testis, ovary, adipose tissue

2. **Generation of reducing equivalents**

   The major role of the pathway is to provide cytoplasmic NADPH for **reductive biosynthesis** of fatty acids, cholesterol and steroids.

3. **Erythrocyte membrane**

   NADPH is required by the RBC to preserve the integrity of RBC membrane.

4. **Lens of eye**

   For preserving the transparency of lens, NADPH is required.

5. **Availability of ribose**

   Ribose and deoxy-ribose are required for DNA/RNA synthesis.

6. **What about ATP?**

   ATP is neither utilised nor produced by the HMP shunt pathway.

Table 9.2. NAD$^+$ and NADP$^+$ are different

NADH is used for reducing reactions in catabolic pathways, e.g., pyruvate to lactate. NADH enters the electron transport chain, and ATP is generated. NADPH is used for reductive biosynthetic reactions, e.g., *de novo* synthesis of fatty acid, synthesis of cholesterol, etc. NADPH will not generate ATP.

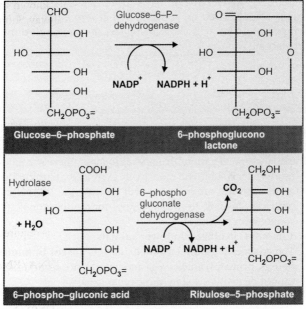

Fig. 9.1. Oxidative phase of HMP shunt pathway

Table 9.3. Glucose-6-phosphate dehydrogenase (GPD) deficiency

It is the most common enzyme deficiency seen in clinical practice. It is an X-linked condition. It will lead to **drug-induced hemolytic anemia.** The deficiency is manifested only when exposed to certain drugs or toxins, e.g. intake of **antimalarial drugs** like primaquin and ingestion of fava beans (**Favism**). Sulpha drugs and furadantin may also precipitate the hemolysis. This is characterised by jaundice and severe anemia. GPD deficient persons will show increased **met-hemoglobin** in circulation.

Table 9.4. Glucuronic acid pathway

It provides **UDP-glucuronic acid,** which is used for:
1. Conjugation of bilirubin and steroids.
2. Conjugation of various drugs so as to make these substance more water soluble and more easily excretable. Barbiturates, antipyrine and aminopyrine will increase the uronic acid pathway, leading to availability of more glucuronate for conjugation purpose.
3. **Vitamin C synthesis in lower animals**
   The enzyme **L-gulonolactone oxidase is absent in human beings**, primates, guinea pigs and bats. Hence ascorbic acid cannot be synthesised by these organisms.

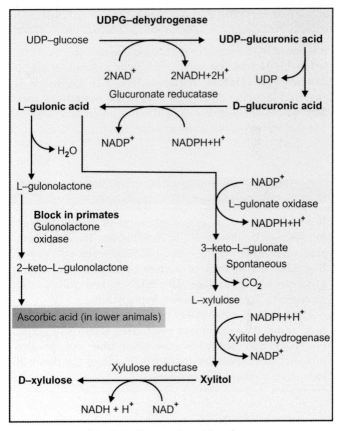

Fig. 9.2. Glucuronic acid pathway

Table 9.5. Essential pentosuria

It is one of the members of the Garrod's tetrad. The incidence is 1 in 2,500 births. It is an inborn error of metabolism due to deficiency of **xylulose reductase** (Fig. 9.2). L-xylulose is excreted in urine and gives a **positive Benedict's test**. Barbiturates, aminopyrine, etc. will induce uronic acid pathway and will increase xylulosuria in such patients.

Table 9.6. Hereditary fructose intolerance (HFI)

It is an autosomal recessive inborn error of metabolism. Incidence of the disease is 1 in 20,000 births. The defect is in aldolase-B; hence fructose-1-phosphate cannot be metabolised (Fig. 9.3). Accumulation of fructose-1-phosphate will inhibit glycogen phosphorylase. This leads to accumulation of glycogen in liver and associated hypoglycemia.

The infants often fail to thrive. Hepatomegaly and jaundice may occur. If liver damage progresses, death will occur. Fructose is also excreted in urine when urine gives positive Benedict's test. Withdrawal of fructose from the diet will immediately relieve the symptoms.

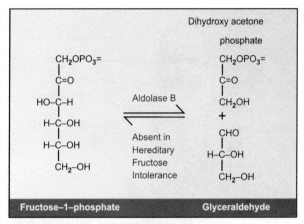

Fig. 9.3. Fructose metabolism

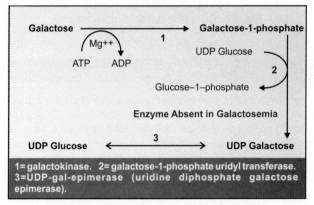

Fig. 9.4. Summary of galactose metabolism

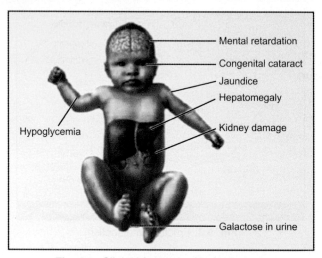

Fig. 9.5. Clinical features of galactosemia

Table 9.7. Galactosemia

There is deficiency of enzyme galactose-1-phosphate uridyl transferase. It is an inborn error of metabolism. Galactose-1-phosphate will accumulate in liver. Hypoglycemia, jaundice, hepatomegaly, severe mental retardation, galactosemia, galactosuria and congenital cataract are seen. Treatment is to give lactose-free diet.

Table 9.8. Features of mucopolysaccharidoses

Name	Deficient enzyme	Clinical findings
Hurler's	L-iduronidase	MR^{+++}; Skeletal deformity^{++}; Corneal opacity^{++}; DS and HS in urine.
Hunter's	Iduronate sulphatase	MR^{+}; Skeletal^{++}; DS and HS in urine
Sanfilippo's	N-acetyl glucosaminidase Heparan sulfatase	MR^{++}; Skeletal^{+}; corneal clouding^{+}; HS in urine
Morquio's	Galactosamine sulfatase, beta-galactosidase	MR^{+}; Skeletal^{+}; epiphysealdysplasia^{+} KS and CS in urine
Scheie's	L-iduronidase	No MR; corneal $^{++}$; DS in urine
Maroteaux-Lamy's	N-acetyl-beta-galactosamino-4-sulfatase	Skeletal deformity^{+++} corneal opacity^{++}; No MR; DS in urine
Sly's	beta-glucuronidase	MR^{+}; DS and HS in urine

MR = mental retardation; CS = chondroitin sulphate;
KS = keratan sulphate;   HS = heparan sulphate;
DS = dermatan sulphate

Table 9.9. Inborn errors associated with carbohydrate metabolism

Name	Incidence 1 out of	Defective enzyme
Glycogen storage disease, Type I (von Gierke's)	**100,000**	**Glucose-6-phosphatase**
Do, type II (Pompe's)	175,000	Lysosomal maltase
Do, type III (Cori's)	125,000	Debranching enzyme
Do, type IV (Andersen's)	1 million	Branching enzyme
Do, type V (McArdle's)	1 million	Muscle phosphorylase
Lactose intolerance		Lactase
Fructose intolerance	**20,000**	**Aldolase B**
Fructosuria	130,000	Fructokinase
Galactosemia	35,000	Gal-1-phosphate-uridyl transferase
Do, variant	40,000	Galactokinase
Essential pentosuria	2,500	Xylulose reductase
PC deficiency	25,000	Pyruvate carboxylase
GPD deficiency	5,000	Glucose-6-phosphate dehydrogenase
HK deficiency		Hexokinase
PK deficiency		Pyruvate kinase
PDH deficiency	250,000	Pyruvate dehydrogenase

Contd...

Contd...

Name	Salient features
Glycogen storage disease, Typel (von Gierke's)	Hepatomegaly, cirrhosis, hypoglycemia, ketosis, **hyperuricemia**
Do, type II (Pompe's)	Generalised glycogen deposit; lysosomal storage disease
Do, type III (Cori's)	Hepatomegaly, cirrhosis
Do, type IV (Andersen's)	Do
Do, type V (McArdle's)	Exercise intolerance
Lactose intolerance	Milk induced diarrhea
Fructose intolerance	**Hypoglycemia,vomiting, hepatomegaly**
Fructosuria	Benign; urine sugar
Galactosemia	**Hypoglycemia ; hepato-megaly; mental** retardation; jaundice; **congenital cataract**
Do, variant	Congenital cataract
Essential pentosuria	Benign
PC deficiency	Mental retardation
GPD deficiency	X-link; Drug-induced hemolytic anemia
HK deficiency	Hemolytic anemia
PK deficiency	Hemolytic anemia
PDH deficiency	Neuronal loss in brain; muscular hypotonia; lactic acidosis

10. Insulin and Diabetes Mellitus

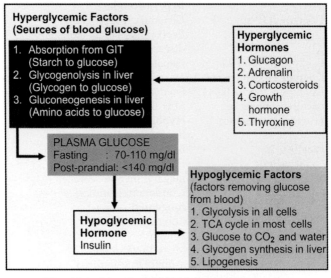

Hyperglycemic Factors
(Sources of blood glucose)

1. Absorption from GIT (Starch to glucose)
2. Glycogenolysis in liver (Glycogen to glucose)
3. Gluconeogenesis in liver (Amino acids to glucose)

Hyperglycemic Hormones
1. Glucagon
2. Adrenalin
3. Corticosteroids
4. Growth hormone
5. Thyroxine

PLASMA GLUCOSE
Fasting   : 70-110 mg/dl
Post-prandial: <140 mg/dl

Hypoglycemic Hormone
Insulin

Hypoglycemic Factors
(factors removing glucose from blood)
1. Glycolysis in all cells
2. TCA cycle in most cells
3. Glucose to CO_2 and water
4. Glycogen synthesis in liver
5. Lipogenesis

Fig. 10.1. Overview of regulation of blood sugar

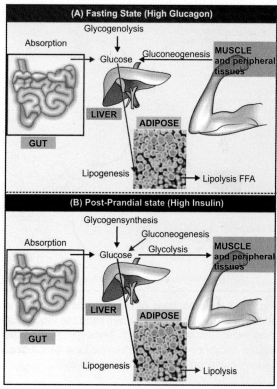

Fig. 10.2. Blood glucose regulation (A) In fasting state, blood glucose level is maintained by glycogenolysis and gluconeo-genesis; adipose tissue releases free fatty acids, (B) In post-prandial state, glucose level is high ——→ = activation ——→ = inhibition

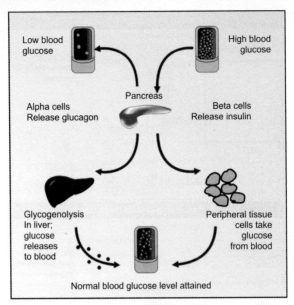

Fig. 10.3. Homeostasis of blood glucose. Combined action of insulin and glucagon will keep the blood sugar level within normal limits. High blood sugar stimulates insulin secretion (green pathway). Low blood sugar causes glucagon secretion (blue pathway)

Table 10.1. Effects of hormones on glucose level

A. **Effect of Insulin (hypoglycemic hormone)**
   1. Lowers blood glucose
   2. Favors glycogen synthesis
   3. Promotes glycolysis
   4. Inhibits gluconeogenesis
B. **Glucagon (hyperglycemic hormone)**
   1. Increases blood glucose
   2. Promotes glycogenolysis
   3. Enhances gluconeogenesis
   4. Depresses glycogen synthesis
   5. Inhibits glycolysis (Details given below).
C. **Cortisol (hyperglycemic hormone)**
   1. Increases blood sugar level
   2. Increases gluconeogenesis
   3. Releases amino acids from the muscle
D. **Adrenaline or Epinephrine (hyperglycemic)**
   1. Increases blood sugar level
   2. Promotes glycogenolysis
   3. Increases gluconeogenesis
   4. Favours uptake of amino acids
E. **Growth Hormone (hyperglycemic)**
   1. Increases blood sugar level
   2. Decreases glycolysis
   3. Mobilises fatty acids from adipose tissue

Table 10.2. Terminology of glucose level

i. Blood sugar analysed at any time of the day, without any prior preparations, is called **random blood sugar**.

ii. Sugar estimated in the early morning, before taking any breakfast (12 hr fasting) is called **fasting blood sugar.**

iii. The test done about 2 hr after a good meal is called **post-prandial blood sugar** (Latin = after food).

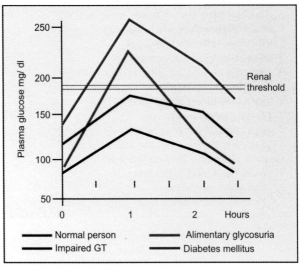

Fig. 10.4. Oral glucose tolerance test (OGTT)

Table 10.3. Oral glucose tolerance test (OGTT)

Indications for OGTT
1. Patient has symptoms suggestive of diabetes mellitus; but fasting blood sugar value is inconclusive (between 110 and 126 mg/dl).
2. During pregnancy, excessive weight gaining is noticed, with a past history of big baby (more than 4 kg) or a past history of miscarriage.

Contraindication for OGTT
GTT has no role in follow-up of diabetes. It is indicated only for the initial diagnosis.

Conducting the glucose tolerance test
1. At about 8 am, a sample of blood is collected in the fasting state. Urine sample is also obtained. This is denoted as the "0" hour sample.
2. **Glucose Load Dose:** The dose is **75 g anhydrous glucose** (82.5 g of glucose monohydrate) in 250-300 ml of water. Dose is adjusted as **1.75 g/kg body weight** in children.
3. **Sample Collection:** In the classical procedure, the blood and urine samples are collected at 1/2 an hour intervals for the next 2½ hours. (Total six samples).
4. But the present WHO recommendation is to collect only the **fasting and 2-hour** post-glucose load samples of blood and urine. This is sometimes called **mini-GTT.**

Table 10.4. The plasma sugar levels in OGTT in normal persons and in diabetic patients

	Normal persons	*Criteria for diagnosing diabetes*	*Criteria for diagnosing IGT*
Fasting	< 110 mg/dl < (6.1mmol/L)	> 126 mg/dl > (7.0 mmol/L)	110 to 126 mg/dl
1 hr (peak) after glucose	< 160 mg/dl < (9 mmol/L)	Not prescribed	Not prescribed
2 hr after glucose	< 140 mg/dl < (7.8 mmol/L)	> 200 mg/dl > (11.1mmol/L)	140 to 199 mg/dl

Table 10.5. Post-glucose load values and post-prandial values are different

Oral glucose tolerance test (OGTT) is always non-physiological, where glucose is administered in a large dose. In this test, the blood sugar reaches a **peak value at 1 hour** and comes back to fasting value by about 2 hour. These are referred to as **post-glucose load** values.

But the **physiological tolerance test** often employed in clinical practice is different. Here fasting blood sample is taken; then patient is asked to take a heavy breakfast (instead of glucose load). The blood sample at 2 hr after the meal is also taken; when the result is referred to as **post-prandial** (Latin, after food) value. In normal persons, the **peak value is at about 2 hr.**

Table 10.6. Causes for abnormal GTT curve

1. Impaired Glucose Tolerance (IGT)
It is otherwise called as **Impaired Glucose Regulation** (IGR). Here blood sugar values are above the normal level, but below the diabetic levels (see Table 10.4). Such persons need careful follow up because IGT progresses to frank diabetes.

2. Gestational Diabetes Mellitus (GDM)
This term is used when carbohydrate intolerance is noticed, for the first time, during a pregnancy. **A known diabetic patient, who becomes pregnant, is not included in this category.** GDM is associated with an increased incidence of **neonatal mortality**.

3. Alimentary Glucosuria
Here the fasting and 2 hr values are normal; but an exaggerated rise in blood glucose following the ingestion of glucose is seen. This is due to an increased rate of absorption of glucose from the intestine. This is seen in patients after a **gastrectomy** or in **hyperthyroidism**.

4. Renal Glucosuria
Normal renal threshold for glucose is 175-180 mg/dl. If blood sugar rises above this, glucose starts to appear in urine. When renal **threshold is lowered**, glucose is excreted in urine. In these cases, the blood sugar levels are within normal limits.

Table 10.7. Reducing substances in urine

Sugars	Noncarbohydrates
Glucose	Homogentisic acid
Fructose	Salicylates
Lactose	Ascorbic acid
Galactose	Glucuronides of drugs
Pentoses	

Table 10.8. Differential diagnosis of reducing substances in urine

1. **Hyperglycemic glucosuria**
   i. *Diabetes mellitus* is the most common cause.
   ii. *Transient glucosuria*. It may occur due to emotional stress, when excessive secretion of cortisol is seen.
   iii. Renal glucosuria, renal threshold is lowered.
   iv. Alimentary glucosuria
2. **Fructose intolerance:** Aldolase-B deficiency
3. **Lactosuria :** It is observed in the urine of normal women during 3rd trimester of pregnancy and during lactation.
4. **Galactosuria :** due to deficiency of galactose-1-phosphate uridyl transferase
5. **Essential Pentosuria:** Excretion of **L-xylulose** in urine due to deficiency of enzyme xylulose reductase.
6. Non-carbohydrate sources are shown in Table 10.7

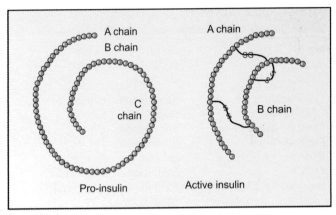

Fig. 10.5. Insulin biosynthesis

Table 10.9. Structure of insulin

i.   Insulin is a protein hormone with 2 polypeptide chains. The A chain has 21 amino acids and B chain has 30 amino acids.

ii. These two chains are joined together by two interchain disulphide bonds, between A7 to B7 and A20 to B19.

iii. There is also an intrachain disulphide link in A chain between 6th and 11th amino acids (Fig. 2.6).

iv. Insulin is synthesised as pro-insulin (Fig.10.5), and later the connecting chain (C chain) is removed.

Table 10.10. Biological effects of insulin

Key enzyme	Action of insulin on the enzyme	Direct effect
Translocase	Stimulation	Glycolysis
Glucokinase	Stimulation	favored
Phospho fructo kinase	Stimulation	Do
Pyruvate kinase	Stimulation	Do
Pyruvate carboxylase	Inhibition	Gluco-
PEPCK	Inhibition	neogenesis
Fructose-1,6-bisphosphatase	Inhibition	depressed
Glucose-6-phosphatase	Inhibition	
Glycogen synthase	Activation	Glycogen deposition
Glycogen phosphorylase	Inactivation	Do
G-6-phosphate dehydrogenase	Stimulation	Generation of NADPH
Acetyl CoA carboxylase	Stimulation	Lipogenesis favored
Glycerol kinase	Stimulation	
Hormone sensitive lipase	Inhibition	Lipolysis inhibited
HMG CoA reductase	Stimulation	Cholesterol synthesis
Transaminases	Inhibition	Catabolism inhibited
RNA polymerase	Protein synthesis	General anabolism

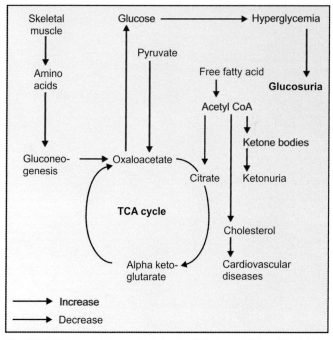

Fig. 10.6. Metabolic derangements in diabetes mellitus. Glycolysis is inhibited; gluconeogenesis is favored. Fat is broken down; FFA is increased; Acetyl CoA is in plenty. This could not be fully utilised in TCA cycle, because availability of oxaloacetate is reduced. So acetyl CoA is shunted to ketone body formation

Section 3

Lipid Metabolism

11. Metabolism of Fatty Acids

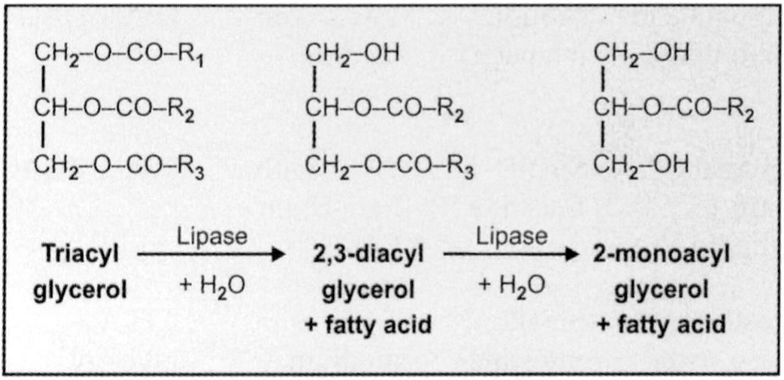

Fig. 11.1. Partial hydrolysis of triglyceride

Table 11.1. Six steps of lipid absorption

1. **Minor digestion** of triacylglycerols in mouth and stomach by lingual (acid-stable) lipase.
2. **Major digestion** of all lipids in the lumen of the duodenum/jejunum by pancreatic lipolytic enzymes.
3. **Bile acid** facilitated formation of mixed micelles.
4. **Passive absorption** of the lipolytic products from the mixed micelle into the intestinal epithelial cell.
5. **Re-esterification** of 2-monoacylglycerol with free fatty acids inside the intestinal enterocyte.
6. **Assembly** of chylomicrons containing Apo B48, triacylglycerols esters and phospholipids **and export** from intestinal cells to the lymphatics.

Table 11.2. Physiologically important lipases

Lipase	Site of action	Preferred substrate	Product(s)
Lingual/acid-stable lipase	Mouth, stomach	TAGs with medium chain FAs	FFA + DAG
Pancreatic lipase + co-lipase	Small intestine	TAGs with long-chain FAs	FFA + 2MAG
Intestinal lipase with bile acids	Small intestine	TAGs with medium chain FAs	3 FFA + glycerol
Phospho-lipase A_2 + bile acids	Small intestine	PLs with unsat. FA on position 2	Unsat FFA lysolecithin
Lipoprotein lipase insulin (+)	Capillary walls	TAGs in chylomicron or VLDL	FFA+ glycerol
Hormone sensitive lipase	Adipose cell	TAG stored in adipose cells	FFA+ glycerol

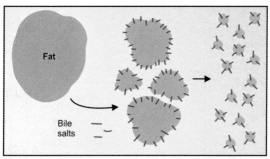

Fig. 11.2. Action of bile salts. The hydrophobic portions of bile salts intercalate into the large aggregated lipid, with the hydrophilic domains remaining at the surface. This leads to breakdown of large aggregates into smaller and smaller droplets. Thus the surface area for action of lipase is increased

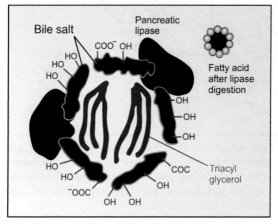

Fig. 11.3. Bile salts and micelle structure

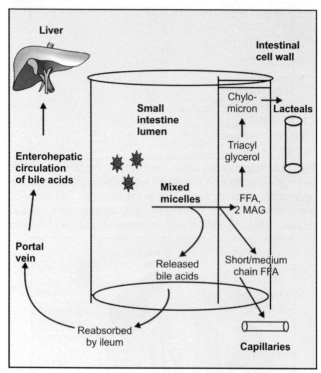

Fig. 11.4. Absorption of fatty acids. Long chain fatty acids are absorbed into the intestinal cell wall, where they are re-esterified, made into chylomicrons and enter into lymphatics. Short chain fatty acids are directly absorbed into blood capillaries. Bile acids are reabsorbed into portal vein

Table 11.3. Abnormalities in absorption of lipids

1. **Defective digestion:** In steatorrhea, daily excretion of fat in feces is more than 6 g per day. It is due to chronic diseases of pancreas. In such cases, unsplit fat is seem in feces.
2. **Defective absorption:** On the other hand, if the absorption alone is defective, most of the fat in feces may be split fat, i.e. fatty acids and monoglycerides. Defective absorption may be due to **obstruction of bile duct.** This again may be due to gallstones, tumors of head of pancreas, enlarged lymph glands, etc. The result is deficiency of bile salts. In such cases, triglycerides with short chain and medium chain fatty acids (SCT and MCT) are digested and absorbed properly, because they do not require micellerisation for absorption.

Table 11.4. Fate of chylomicrons

i. The absorbed (exogenous) triglycerides are transported in blood as **chylomicrons**. They are taken up by adipose tissue and liver.
ii. Triglycerides in adipose tissue are lysed to produce **free fatty acids**. In the blood, they are transported, complexed with albumin.
iii. Free fatty acids are taken up by the cells, and are then oxidised to get energy.

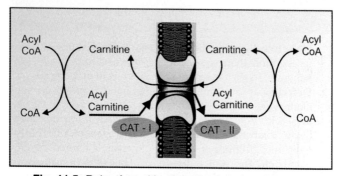

Fig. 11.5. Role of carnitine in transport of acyl groups.
CAT = Carnitine acyl transferase

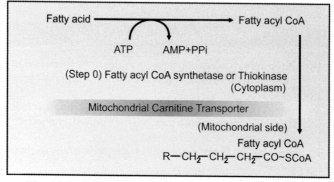

Fig. 11.6. Step 0 (Preparatory step) of beta oxidation of
fatty acids

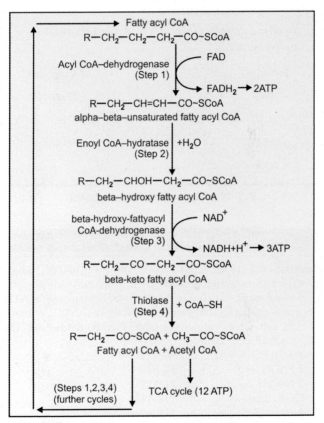

Fig. 11.7. Beta oxidation of fatty acids. Important to remember that the first step is FAD dependent and the third step is NAD^+ dependent

When one molecule of palmitate undergoes beta-oxidation, the net reaction is:

Palmitoyl CoA ⎯⎤   ⎡⎯ 8 Acetyl CoA

+ 7FAD   + 7FADH$_2$

+ 7NAD$^+$ ⎬ ⎯⎯⎯→ ⎨   + 7NADH

+ 7H$_2$O   + 7H$^+$

+ 7HSCoA ⎯⎦

Fig. 11.8. Summary of beta oxidation

Table 11.5. Energetics of beta oxidation

Palmitic acid (16 C) needs 7 cycles of beta oxidation, which give rise to 8 molecules of acetyl CoA. Every molecule of acetyl CoA when oxidised in the TCA cycle gives 12 molecules of ATP. Each molecule of FADH$_2$ produces 2 molecules of ATP and each NADH generates 3 molecules of ATP when oxidised in the electron transport chain.

8 acetyl CoA × 12	=	96 ATP
7 FADH$_2$ × 2	=	14 ATP
7 NADH × 3	=	21 ATP
Gross total	=	131 ATP
Net yield	**=**	**131 minus 2 = 129 ATP**

(In the initial activation reaction, the equivalent of 2 high energy bonds are utilised).

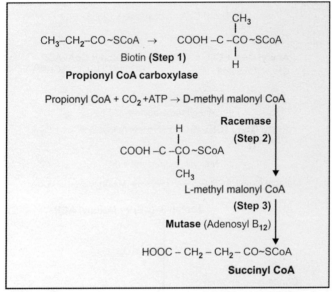

Fig. 11.9. Metabolism of propionyl CoA. Ordinary fatty acids are cleaved to acetyl CoA units which on entering the Krebs cycle are completely oxidised to CO_2, and hence fatty acids cannot be used for gluconeogenesis. However propionate is entering into the citric acid cycle at a point after the CO_2 elimination steps, so propionate can be channeled to gluconeogenesis. Thus 3 carbon units from odd carbon fatty acids are gluconeogenic. Cow's milk contains significant quantity of odd chain fatty acids

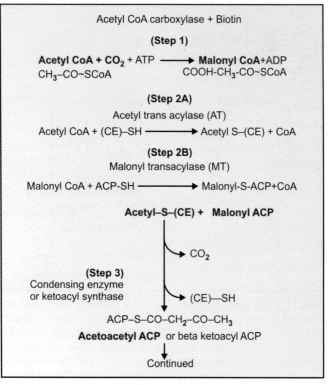

Fig. 11.10.

Contd...

Contd...

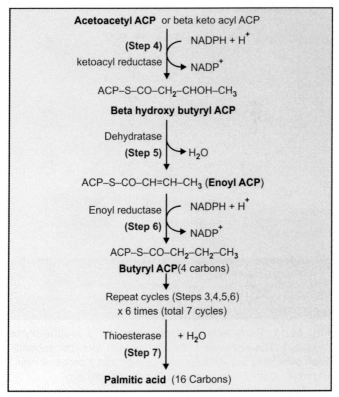

Fig. 11.10. *De novo* synthesis of fatty acid (Lynen cycle).
Steps 4 and 6 utilise NADPH

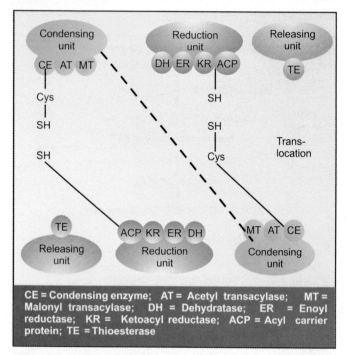

CE = Condensing enzyme; AT = Acetyl transacylase; MT = Malonyl transacylase; DH = Dehydratase; ER = Enoyl reductase; KR = Ketoacyl reductase; ACP = Acyl carrier protein; TE = Thioesterase

Fig. 11.11. Fatty acid synthase complex (a multi-enzyme complex). The enzymes form a dimer with identical subunits. Each subunit of the complex is organised into 3 domains with 7 enzymes. Dotted line represents functional division

Table 11.6. Succinyl CoA pool

Succinate donors		Succinate utilisation
Krebs cycle →		→ Krebs cycle → glucose
Odd no. FA →	Succi-	→ Porphyrin synthesis
Propionic acid →	nyl	→ Detoxification
Threonine →	CoA	→ Ketone body activation
Valine →	pool	→ Activation of small
Isoleucine →		chain fatty acids

Table 11.7. Difference in the two pathways

	Beta-oxidation	Fatty acid synthesis
Site	Mitochondria	Cytoplasm
Intermediates	Present as CoA derivatives	Covalently linked SH group of ACP
Enzymes	Present as independent proteins	Multi-enzyme complex
Sequential units	2 carbon units split off as acetyl CoA	2 carbon units added, as 3 C malonyl CoA
Co-enzymes	NAD^+ and FAD are reduced	NADPH used as reducing power

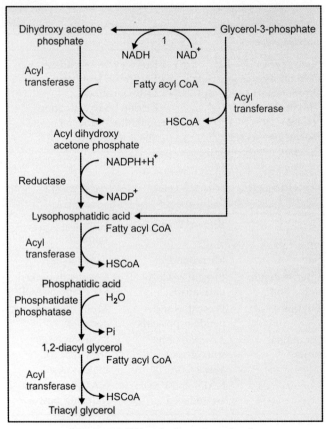

Fig. 11.12. Triacyl glycerol synthesis.
1 = glycerol-3-phosphate dehydrogenase

Table 11.8. Regulation of fatty acid and triacyl glycerol synthesis

1. Acetyl CoA carboxylase

It is the **key enzyme**; citrate activates this enzyme. The citrate level is high only when both acetyl CoA and ATP are abundant. Fatty acid synthesis decreases when glucose level is low. The enzyme is inhibited by palmitoyl CoA, the end product.

2. Insulin favors lipogenesis

Insulin enhances the uptake of glucose by adipocytes and increases the activity of pyruvate dehydrogenase, acetyl CoA carboxylase and glycerol phosphate acyl transferase. Insulin also depresses the hormone sensitive lipase. Insulin also causes inhibition of **hormone sensitive lipase**, and so lipolysis is decreased.

3. Glucagon inhibits lipogenesis

Glucagon and epinephrine inactivate the acetyl CoA carboxylase.

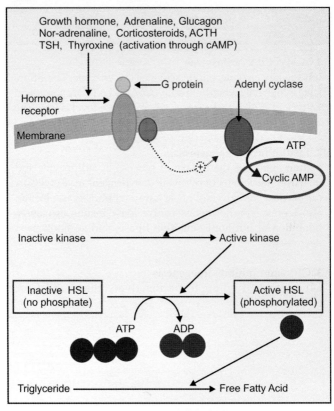

Fig. 11.13. Cascade activation of HSL
(hormone sensitive lipase)

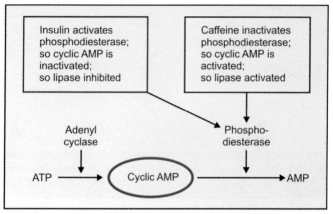

Fig. 11.14. Production and inactivation of cyclic AMP and action of insulin

Table 11.9. Role of liver in fat metabolism

1. Secretion of bile salts.
2. Synthesis of fatty acid, triacyl glycerol and phospholipids.
3. Oxidation of fatty acids.
4. Production of lipoproteins.
5. Production of ketone bodies.
6. Synthesis and excretion of cholesterol.

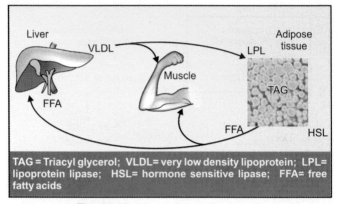

TAG = Triacyl glycerol; VLDL= very low density lipoprotein; LPL= lipoprotein lipase; HSL= hormone sensitive lipase; FFA= free fatty acids

Fig. 11.15. Liver-adipose tissue axis

Table 11.10. Lipotropic factors

They can afford protection against the development of fatty liver.

1. **Choline**: Feeding of choline has been able to reverse fatty changes in animals.
2. **Lecithin and methionine.** They help in synthesis of apoprotein and choline formation.
3. **Vitamin E and selenium** give protection due to their antioxidant effect.
4. **Omega 3 fatty acids** present in marine oils.

Table 11.11. Causes of fatty liver

A. Causes of fat deposition in liver
1. Mobilisation of NEFA from adipose tissue.
2. More synthesis of fatty acid from glucose.

B. Reduced removal of fat from liver
3. Toxic injury to liver. Secretion of VLDL needs synthesis of apo B-100 and apo C.
4. Decreased oxidation of fat by hepatic cells.

An increase in factors (1) and (2) or a decrease in factors (3) and (4) will cause excessive accumulation, leading to fatty liver.

1. Excessive mobilisation of fat
The capacity of liver to take up the fatty acids from blood far exceeds its capacity for excretion as VLDL. So fatty liver can occur in **diabetes mellitus and starvation** due to increased lipolysis in adipose tissue.

2. Excess calorie intake

3. Toxic injury to liver
Hepatitis B virus infection reduces the function of hepatic cells.

4. Alcoholism
It is the most common cause of fatty liver and cirrhosis. Fatty acid accumulates leading to TAG deposits in liver.

5. Fatty liver progresses to cirrhosis.

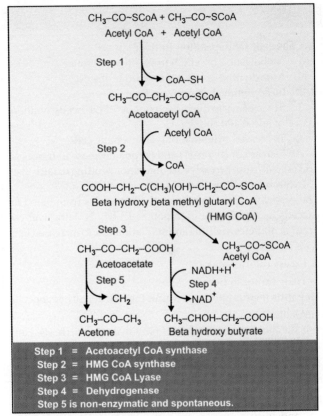

Fig. 11.16. Ketone body formation (ketogenesis)

Table 11.12. Ketolysis

i. The ketone bodies are formed in the liver; but they are utilised by **extrahepatic tissues**. The heart muscle and renal cortex prefer the ketone bodies to glucose as fuel. Tissues like skeletal muscle and brain can also utilise the ketone bodies as alternate sources of energy, if glucose is not available.

ii. Acetoacetate is activated to acetoacetyl CoA by **thiophorase** enzyme.

$$\text{Acetoacetate} + \text{Succinyl CoA} \xrightarrow{\text{Thiophorase}} \text{Acetoacetyl CoA} + \text{Succinate}$$

Then acetoacetyl CoA enters the beta oxidation pathway to produce energy.

Table 11.13. Ketosis

i. Normally the rate of synthesis of ketone bodies by the liver is such that they can be easily metabolised by the extrahepatic tissues.

ii. But when the rate of synthesis exceeds the ability of extrahepatic tissues to utilise them, there will be accumulation of ketone bodies in blood.

iii. This leads to **ketonemia**, excretion in urine (**ketonuria**) and smell of **acetone** in breath. All these three constitute the condition known as **ketosis**.

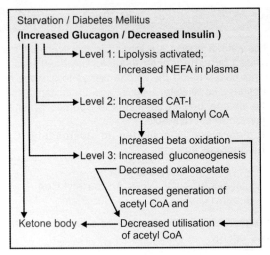

Fig. 11.17: Summary of ketosis

Table 11.14. Consequences of ketosis

1. **Metabolic acidosis.** Acetoacetate and beta-hydroxy butyrate are acids. When they accumulate, metabolic acidosis results.
2. **Buffers.** The plasma bicarbonate is used up for buffering of these acids.
3. **Kussmaul's respiration.** Hyperventillation leads to typical acidotic breathing.
4. **Smell of acetone** in patient's breath.
5. **Coma,** hypokalemia, dehydration and death.

12. Cholesterol Metabolism

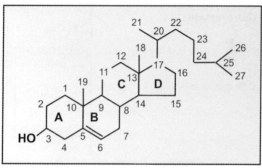

Fig. 12.1: Structure of cholesterol

Table 12.1. Significance of cholesterol

1. Heart diseases: The level of cholesterol in blood is related to the development of atherosclerosis.
2. Cell membranes: Cholesterol has a modulating effect on the fluid state of the membrane.
3. Nerve conduction: Cholesterol is used to insulate nerve fibers.
4. Bile acids and bile salts: The 24 carbon bile acids are derived from cholesterol.
5. Steroid hormones: Cholesterol is the precursor of 21carbon glucocorticods, 19 carbon androgens and 18 carbon estrogens.
6. Vitamin D: It is synthesised from cholesterol.

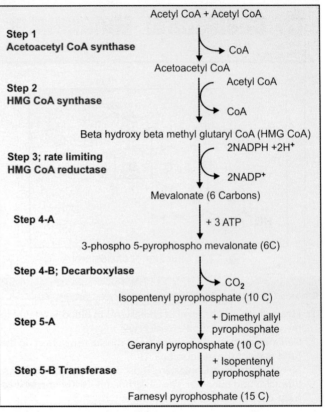

Fig. 12.2.

Contd...

Contd...

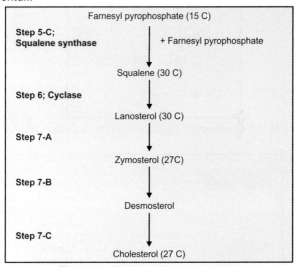

Fig.12.2. Cholesterol biosynthesis

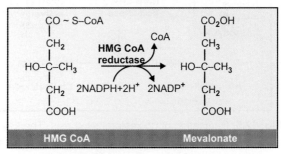

Fig.12.3. Step 3 of cholesterol synthesis

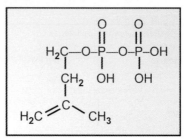

Fig. 12.4. Isopentenyl pyrophosphate; 5 carbon unit

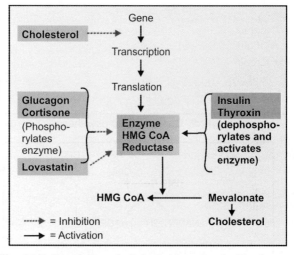

Fig. 12.5. Regulation of cholesterol synthesis. The key enzyme is HMG CoA reductase

Table 12.2. Summary of cholesterol metabolism

1. Incorporated into cell membranes.
2. Metabolised to steroid hormones, especially in adrenal cortex and gonads.
3. Re-esterified and stored. The enzyme ACAT (acyl cholesterol acyl transferase) utilises monounsaturated fatty acids for re-esterification and deposit cholesterol esters in the cell.
4. Expulsion of cholesterol from the cell, esterification with poly-unsaturated fatty acids (PUFA) by the action of LCAT (lecithin cholesterol acyl transferase) and transported by HDL and finally excreted through liver.
5. Excretion of cholesterol: Average diet contains about 300 mg of cholesterol per day. Body synthesise about 700 mg of cholesterol per day. Out of this total 1000 mg, about 500 mg of cholesterol is excreted through bile. This cholesterol is partly reabsorbed from intestines. Vegetables contain plant sterols which inhibit the re-absorption of cholesterol. The unabsorbed portion is acted upon by intestinal bacteria to form cholestanol and coprostanol. These are excreted (fecal sterols). Another 500 mg of cholesterol is converted to bile acids, which are excreted as bile salts.

Table 12.3. Plasma lipid profile (normal values)

Analyte	Normal value
Total plasma lipids	400-600 mg/dl
Total cholesterol	**150-200 mg/dl**
HDL cholesterol, male	30-60 mg/dl
HDL cholesterol, female	35-75 mg/dl
LDL cholesterol, 30-39 yr	**80-130 mg/dl**
Triglycerides, male	**50-150 mg/dl**
Triglycerides, female	**40-150 mg/dl**
Phospholipids	150-200 mg/dl
Free fatty acids(FFA)(NEFA)	10-20 mg/dl

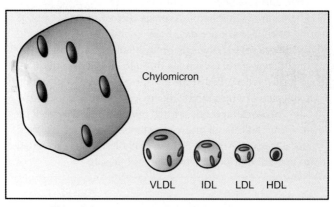

Fig. 12.6. Comparison of sizes of lipoproteins

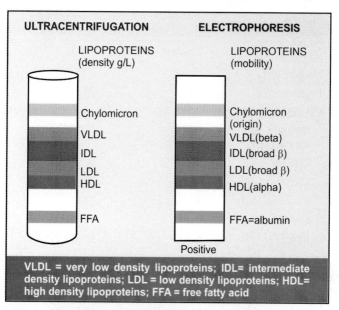

Fig. 12.7. Comparison of electrophoretic and ultracentrifuge patterns of lipoproteins

Table 12.4. Characteristics of lipoproteins

	Chylomicron	VLDL
Density g/L	< 0.95	0.95-1.006
Diameter (nm)	500	70
Electrophoretic mobility	origin	pre-beta
% composition		
Protein	2	10
TAG	83	50
Phospholipids	7	18
Cholesterol	8	22
FFA	0	0
Apoproteins	A, B-48, C-II, E	B-100, C-II, E
Transport function	TAG From gut to muscle and adipose tissue	TAG From liver to muscle

Contd...

Contd...

	LDL	HDL	FFA (*)
Density g/L	1.019 to 1.063	1.063 to 1.121	1.28 to 1.3
Diameter (nm)	25	15	—
Electrophoretic mobility	beta	alpha	albumin
% composition			
Protein	22	30-60	99
TAG	10	8	0
Phospholipids	22	20-30	0
Cholesterol	46	10-30	0
FFA	0	0	1
Apoproteins	B-100	A-I, C, E	Albumin
Transport function	Cholesterol from liver to heart and liver	Cholesterol from heart to liver	FFA from adipose to muscle

(*) Free fatty acids are not generally included in the lipoproteins. They are seen in circulation, weakly bound to albumin

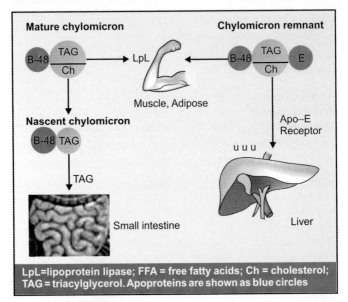

Fig. 12.8. Metabolism of chylomicrons. Chylomicrons are the transport form of dietary triglycerides from intestines to the adipose tissue for storage; and to muscle or heart for their energy needs. Chylomicron remnants containing apo E are taken up by hepatic cells by receptor mediated endocytosis. Apo E binds the hepatic receptors

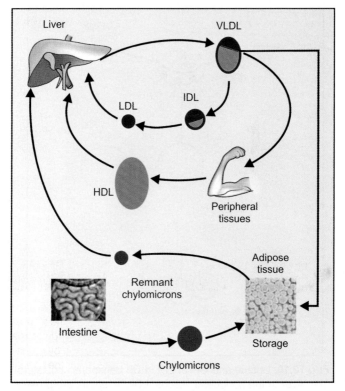

Fig. 12.9. Summary of lipoprotein metabolism. This conversion of VLDL to IDL and then to LDL by losing the triglyceride, is referred to as lipoprotein cascade pathway

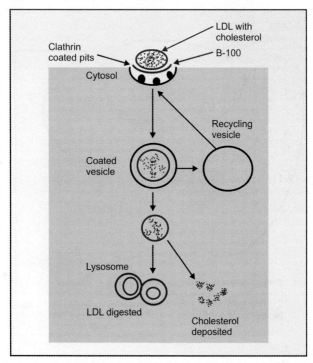

Fig. 12.10. Uptake and fate of LDL. LDL transports cholesterol from liver to the peripheral tissues. LDL receptors are present on all cells but most abundant in hepatic cells. LDL receptors are located in specialised regions called clathrin-coated pits. When the apo B-100 binds to the receptor, the receptor-LDL complex is internalised by endocytosis. These vesicles would fuse with lysosomes

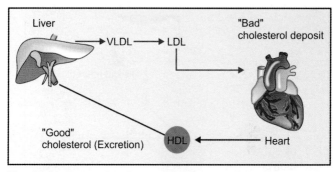

Fig. 12.11. Forward and reverse transport of cholesterol. LDL infiltrates through arterial walls, leading to atherosclerosis and myocardial infarction . So the LDL (low density lipoprotein) variety is called "bad cholesterol" or as "Lethally Dangerous Lipoprotein"

Table 12.5. Lp(a) and apo-A are different

Apo-A is a constituent of HDL. This "A" is always written in capital letters. It is seen in all persons. It is anti-atherogenic.

Lp(a) is seen only in some persons. When present, it is associated with LDL. This "a" is always written in small letters. It is highly atherogenic and connected with heart attacks in younger age group. Lp(a) has significant homology with **plasminogen**. So it interferes with plasminogen activation and impairs fibrinolysis, leading to thrombosis.

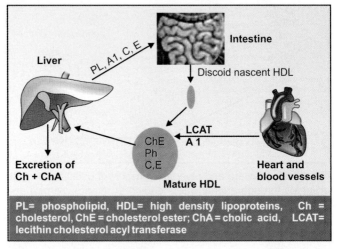

PL= phospholipid, HDL= high density lipoproteins, Ch = cholesterol, ChE = cholesterol ester; ChA = cholic acid, LCAT= lecithin cholesterol acyl transferase

Fig. 12.12. HDL metabolism. The nascent HDL in plasma are discoid in shape. The apo A-I of HDL activates LCAT (lecithin cholesterol acyl transferase) present in the plasma. The LCAT then binds to the HDL disc. The second carbon of lecithin contains one molecule of polyunsaturated fatty acid (PUFA). It is transferred to cholesterol to form cholesterol ester. The esterified cholesterol moves into the interior of the HDL disc. Mature HDL spheres are taken up by liver cells. HDL is the main transport form of cholesterol from peripheral tissue to liver

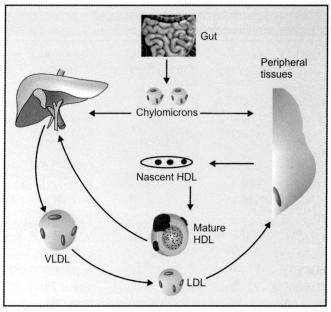

Fig. 12.13. Summary of lipoprotein metabolism. The level of HDL in serum is inversely related to the incidence of myocardial infarction. As it is "anti-atherogenic" or "protective" in nature, HDL is known as "good cholesterol" or "Highly Desirable Lipoprotein" in common parlance. It is convenient to remember that "H" in HDL stands for "Healthy"

Table 12.6. Frederickson's classification

Types	Lipo protein fraction elevated	Chole-sterol level	TAG level	Appearance of plasma after 24 hr
Type I	Chylo-microns	↑	↑↑	Creamy layer over clear plasma
Type II A	LDL	↑↑	N	Clear
Type II B	LDL and VLDL	↑↑	↑	Slightly cloudy
Type III	Broad beta-VLDL and chylomicrons	↑↑	↑	Cloudy
Type IV	VLDL	↑	↑↑	Cloudy or milky
Type V	VLDL chylo microns	N	↑↑	Creamy layer over milky plasma

Contd...

Contd...

Types	Metabolic defect	Features
Type I	Lipoprotein lipase	Eruptive xanthoma; hepatomegaly; Pain abdomen
Type II A	LDL Receptor defect; Apo B ↑	Atherosclerosis, coronary artery disease, Tuberous xanthoma
Type II B	Apo B ↑	Corneal arcus
Type III	Abnormal Apo E; Apo CII ↑	Palmar xanthoma High incidence of vascular disease
Type IV	Over pro-duction of VLDL; Apo CII ↑	Associated with diabetes mellitus, ischemic heart disease, obesity.
Type V	Secondary to other causes	Ischemic heart diseases

Table 12.7. Secondary hyperlipemias

	Serum cholesterol	Serum triglyceride
Diabetes	Increased	Increased
Nephrotic syndrome	Increased	Increased
Hypothyroidism	Increased	Increased
Biliary obstruction	Increased	Normal
Pregnancy	Normal	Increased
Alcoholism	Normal	Increased
Oral contraceptives	Normal	Increased

Table 12.8. Serum cholesterol increased in

1. Coronary artery disease and atherosclerosis.
2. Familial hyperlipoproteinemias.
3. Diabetes mellitus: Acetyl CoA pool is increased.
4. Obstructive jaundice: The excretion of cholesterol through bile is blocked.
5. Hypothyroidism: The receptors for HDL on liver cells are decreased, so excretion is not effective.
6. Nephrotic syndrome: Albumin is lost through urine, globulins are increased as a compensatory mechanism. So, apolipoproteins are increased, and then cholesterol is correspondingly increased.

Table 12.9. Risk factors for atherosclerosis

1. **Serum cholesterol level:**
   In normal persons, cholesterol level varies from 150 to 200 mg/dl. It should be preferably **below 180 mg/dl.** Values around 220 mg/dl will have moderate risk and values above 240 mg/dl will need active treatment.

2. **LDL-cholesterol level:**
   It is "bad" choldestrol. Blood levels **under 130 mg/dl** are desirable. Levels between 130 and 159 are borderline; while above 160 mg/dl carry definite risk.

3. **HDL-cholesterol level:**
   HDL level **above 60 mg/dl** protects against heart disease. Hence HDL is **"good"** cholesterol. A level below 40 mg/dl increases the risk of CAD. If the ratio of total cholesterol/ HDL is more than 3.5, it is dangerous. Similarly, LDL : HDL ratio more than 2.5 is also deleterious.

4. **Apoprotein levels and ratios:**
   Apo A-I is a measure of HDL-cholesterol and apo B measures LDL-cholesterol. Ratio of **Apo B: A-I** is the most reliable index. The ratio of 0.4 is very good; the ratio 1.4 has the highest risk of cardiovascular accidents.

Contd...

Contd...

5. Lp(a):
Levels more than 30 mg/dl increase the risk 3 times; and when increased Lp(a) is associated with increased LDL, the risk is increased 6 times.

6. Cigarette smoke:
Nicotine also causes transient constriction of coronary and carotid arteries.

7. Hypertension:
Systolic blood pressure more than 160 further increases the risk of CAD. An increase in 10 mm of BP will reduce life expectancy by 10 years.

8. Diabetes mellitus:
In the absence of insulin, hormone sensitive lipase is activated, more free fatty acids are formed, these are catabolised to produce acetyl CoA. These cannot be readily utilised, and it is channelled to cholesterol synthesis.

9. Serum triglyceride:
Normal level is 50-150 mg/dl. Blood level **more than 150 mg/dl** is injurious to health.

10. Obesity and sedentary lifestyle
People with "apple type" obesity are more prone to get myocardial infarction.

Table 12.10. Prevention of atherosclerosis

1. Reduce dietary cholesterol
The aim is to reduce total cholesterol below 180 mg/dl; to decrease LDL-cholesterol below 130 mg/dl and to keep HDL-cholesterol above 35 mg/dl. Cholesterol in the diet should be kept less than 200 mg per day. Eggs and meat contain high cholesterol. One egg yolk contains about 500 mg of cholesterol. One double omelet increases the blood cholesterol, 15 mg more than the original level.

2. Vegetable oils and PUFA
Vegetable oils (e.g. sunflower oil) and fish oils contain polyunsaturated fatty acids (PUFA). They are required for the esterification and final excretion of cholesterol. So PUFA is helpful to reduce cholesterol level in blood. **Omega-3 fatty acids** from fish oils reduce LDL and decrease the risk of CAD.

3. Moderation in fat intake
The accepted standard is that about 20% of total calories may be obtained from fat, out of which about one-third from saturated, another one-third from monounsaturated and the rest one-third from polyunsaturated fatty acids. The recommended daily allowance will be about **20-25 g of oils** and about 2-3 g of PUFA per day for a normal adult.

Contd...

Contd...

4. Green leafy vegetables

Due to their **high fibre content,** leafy vegetables will increase the motility of bowels and reduce reabsorption of bile salts. Vegetables also contain plant sterols (**sitosterol**) which decrease the absorption of cholesterol. About 400 g/day of fruit and vegetables are desired.

5. Avoid sucrose and cigarette

Sucrose will raise plasma triglycerides. High carbohydrate diet, especially sucrose, should be avoided by patients with hypercholesterolemia.

6. Exercise

Regular moderate exercise will lower LDL (bad cholesterol) and raise HDL (good cholesterol) levels in blood. It will also reduce obesity.

7. Hypolipidemic drugs

  i. **Bile acid binding resins** (cholestyramine and cholestipol) decrease the reabsorption of bile acids.

 ii. **HMG CoA reductase inhibitors** ("statins"): Atarvostatin and Simvostatin are popular drugs in this group. They are very effective in reducing the cholesterol level and decreasing the incidence of CAD.

13. PUFA and Prostaglandins

Table 13.1. Fatty acids

Fatty acids having carbon atoms 4 to 6 are called **small chain fatty acids** (SCFA); those with 8 to 14 carbon atoms are known as **medium chain fatty acids** (MCFA); those with 16 to 18 carbon atoms are **long chain fatty acids** (LCFA); and those carrying 20 or more carbon atoms are named as **very long chain fatty acids** (VLCFA) (Table 8.1). The important **polyunsaturated fatty acids (PUFA)** are:

1. **Linoleic acid** (18 C, 2 double bonds)
2. **Linolenic acid** (18 C, 3 double bonds)
3. **Arachidonic acid** (20 C, 4 double bonds)

Significance of PUFA are:

1. PUFAs are used for esterification and excretion of **cholesterol**.
2. They are nutritionally essential; and are called **Essential Fatty Acids**.
3. **Prostaglandins**, thromboxanes and leukotrienes are produced from arachidonic acid.
4. They are components of **membranes**. As PUFAs are easily liable to undergo peroxidation, the membranes containing PUFAs are more prone for damage by free radicals.

Table 13.2. Differences in metabolisms of SCFA and LCFA containing triglycerides

	Small and medium chain fatty acids	Long chain fatty acids
Examples	**Butyric acid (C4) in butter; lauric acid (C12) in coconut oil**	**Palmitic acid (C16) and Stearic acid (C18) in vegetable oils and animal fats**
Digestion in stomach	Hydrolysed	Not hydrolysed
Pancreatic lipase	Not necessary	Essential
Bile salts	Not necessary	Absolutely essential
Inside intestinal cells	TAG is hydrolysed to form fatty acids	Free fatty acids are re-esterified to form TAG
Absorbed	**Directly to blood**	**To lymphatics**
Absorbed as	Free fatty acid carried by albumin	TAG, carried by chylomicrons
Immediate fate	Oxidised by peripheral cells	Deposited in the adipose tissue
Carnitine	Not required	Required for oxidation
Clinical application	**No effect on atherosclerosis**	**Hypercholesterolemia and atherosclerosis**

Table 13.3. Modified beta-oxidation of MUFA

The oxidation of unsaturated fatty acids proceeds as in the case of saturated fatty acids, till the double bond is reached. Thus palmitoleic acid (16 C monounsaturated) undergoes 3 cycles of beta-oxidation to yield Δ 3-cis enoyl CoA with 10 carbon atoms. Here the double bond is cis type; the dehydrogenase cannot act on that bond. Therefore, an **isomerase** makes the cis Δ 3 double bond to Δ 2-trans double bond. The double bond between 3rd and 4th carbon atoms is shifted to between 2nd and 3rd carbon atoms. It will then undergo 2nd, 3rd and 4th step reactions of beta oxidation. So in this cycle the **FAD dependent dehydrogenation** (1st step in Fig.11.7) **is not needed.** The energy yield from one molecule of palmitoleic acid will be:

8 acetyl CoA	$\times$ 12	= 96 ATP
6 FADH$_2$	$\times$ 2	= 12 ATP
7 NADH	$\times$ 3	= 21 ATP
Total		= 129 ATP
Net yield	**= 129 minus 2**	**= 127 ATP**

Palmitic acid, a saturated acid, will yield a net 129 ATPs (Table 11.5). Thus in the case of unsaturated fatty acids, the **energy yield is less by 2 ATP molecules per double bond,** because the FAD dependent dehydrogenation (step 1 of beta-oxidation) does not occur at the double bond.

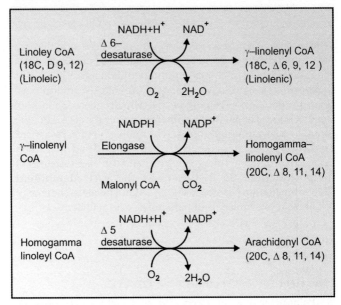

Fig. 13.1. Desaturation and elongation of linoleic acid to arachidonic acid. Linoleic acid ($\omega6$, 18C, $\Delta9,12$) and linolenic acid ($\omega3$, 18C, $\Delta9,12,15$) are the only fatty acids which cannot be synthesised in the body. Hence they are essential fatty acids. Arachidonic acid can be formed, if the dietary supply of linoleic acid is sufficient

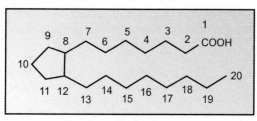

Fig. 13.2. Prostanoic acid

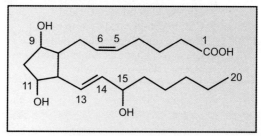

Fig. 13.3. Prostaglandin F$_2$

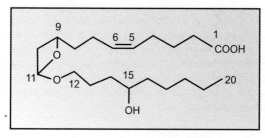

Fig. 13.4. Thromboxane A$_2$

Table 13.4. Salient features of prostaglandins

Name	Substituent groups
PGA	Keto group at C9; double bond C10 and 11
PGB	Keto group at C9; double bond C8 and 12
PGD	OH group at C9; keto group at C11
PGE	Keto group at C9; OH group at C11
PGF	OH groups at C9 and C11
PGG	Two oxygen atoms, interconnected to each other, and bonded at C9 and C11; hydro-peroxide group at C15
PGI	Double ring. Oxygen attached to C6 and C9, to form another 5-membered ring Hence called prostacyclin

Table 13.5. Mechanism of action of aspirin

Aspirin irreversibly acetylates and so inhibits **cyclo-oxygenase**. Platelets cannot regenerate cyclo-oxygenase and so thromboxane A is not formed in platelets. Hence there is decreased platelet aggregation. Therefore, aspirin is useful in prevention of **heart attacks**. By inhibiting cyclo-oxygenase, aspirin also reduces PGI_2; but endothelial cells after a few hours will resynthesise cyclo-oxygenase. So aspirin completely blocks TXA_2, but only partially inhibits PGI_2. **Paracetamol** is a reversible inhibitor.

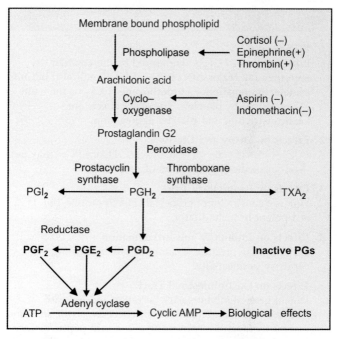

Fig. 13.5. Synthesis and action of prostaglandins

Table 13.6. Biological actions of prostaglandins

1. **Effects on CVS**
   Prostacyclin or PGI_2 is synthesised by the vascular endothelium. Major effect is **vasodilatation**. It also **inhibits platelet aggregation**. **Thromboxane** (TXA_2) is the main PG produced by platelets. The major effects are **vasoconstriction** and **platelet aggregation**.

2. **Effects on Ovary and Uterus**
   PGF_2 stimulates the uterine muscles. Hence PGF_2 may be used for medical **termination of pregnancy**.

3. **Effects on Respiratory Tract**
   PGF is a constrictor of bronchial smooth muscle; but PGE is a potent **bronchodilator**.

4. **Effects on Immunity and Inflammation**
   PGE_2 and D_2 produce inflammation by increasing capillary permeability.

5. **Effects on Gastrointestinal Tract**
   PGs in general inhibit gastric secretion and increase intestinal motility. The inhibitory effect on gastric secretion is used therapeutically in treatment of **acid peptic disease**.

6. **Metabolic Effects**
   Prostaglandin E_2 decreases lipolysis, increases calcium mobilisation from bone and glycogen synthesis. PGE_2 causes fever and stupor.

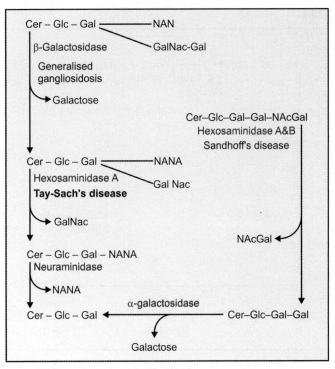

Fig. 13.6.

Contd...

Contd...

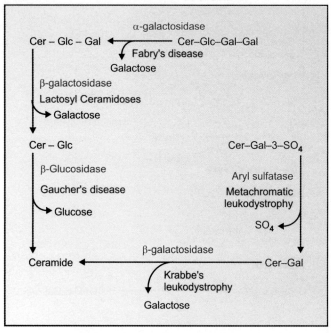

Fig. 13.6. Summary of lipid storage diseases

Table 13.7. Sphingolipidoses or lipid storage diseases

Diseases	Enzyme defects	Lipid accumulating
Gaucher's disease	beta-glucosidase	Gluco cerebroside
Niemann-Pick disease	Sphingo-myelinase	Sphingomyelin
Krabbe's leuko-dystrophy	beta-galactosidase	Galacto-cerebroside
Metachroma-tic leuko-dystrophy	Sulfatide sulfatase	Sulfogalacto-cerebroside
Fabry's disease	alpha-galactosidase	Ceramide trihexoside
Tay-Sach's disease	Hexosaminidase A	Ganglioside (GM_2)
Generalised gangliosidoses	beta-galactosidase (GM_1)	Ganglioside
Lactosyl ceramidoses	beta-galactosidase	Lactosyl ceramide
Sandhoff's disease	Hexosaminidase A and B	Globoside

Section 4

Amino Acid Metabolism

14. General Metabolism of Amino Acids

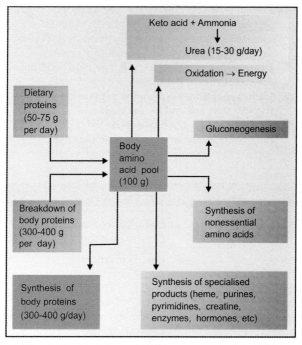

Fig. 14.1. Overview of metabolism of amino acids

Table 14.1. Proteolytic enzymes

Proteolytic enzymes are secreted as inactive **zymogens** which are converted to their active form in the intestinal lumen. This would prevent autodigestion of the secretory acini.

1. **Endopeptidases.** They act on peptide bonds inside the protein molecule, so that the protein becomes successively smaller and smaller units, e.g. Pepsin, Trypsin, Chymotrypsin, and Elastase.
2. **Exopeptidases,** which act at the peptide bond only at the end region of the chain.
2-A. **Carboxypeptidase** acts on the peptide bond only at the carboxy terminal end on the chain.
2-B. **Amino peptidase,** which acts on the peptide bond only at the amino terminal end on the chain.

Table 14.2. Specificity of proteolytic enzymes

Enzymes	Hydrolysis of bonds formed by carboxyl groups of
Pepsin	Phe, Tyr, Trp, Met
Trypsin	Arg, Lys
Chymotrypsin	Phe, Tyr, Trp, Val, Leu
Elastase	Ala, Gly, Ser
Carboxypeptidase A	C-terminal aromatic amino acid
Carboxypeptidase B	C-terminal basic amino acid

Table 14.3. Digestion of proteins

Locations	Agents	Outcomes
Stomach	Acid (HCl)	Denaturation
Stomach	Pepsin	Large peptides + some free amino acids
Small intestine lumen (pancreatic enzymes)	Trypsin chymotrypsin elastase carboxy-peptidases	Oligopeptides (2-10 amino acids) + free amino acids
Brush border surface of intestine	Endopepti-dases and aminopepti-dases	Di- or tri-peptides + free amino acids
Epithelial cell cytoplasm	Dipeptidases and tripeptidases	Free amino acids

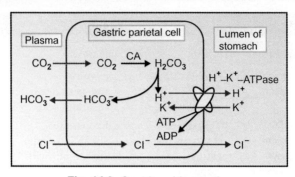

Fig. 14.2. Gastric acid secretion

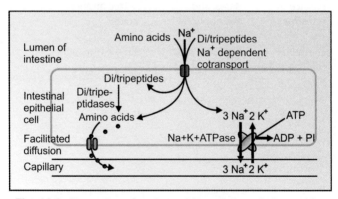

Fig. 14.3. Absorption of amino acids and di- and tri- peptides from the intestinal lumen

Table 14.4. Absorption of amino acids

The absorption of amino acids occurs mainly in the small intestine. It is an energy requiring process. These transport systems are carrier mediated and or ATP sodium dependent symport systems. There are **5 different carriers** for:
1. Neutral amino acids (Alanine, Valine, Leucine, Methionine, Phenylalanine, Tyrosine, Ile)
2. Basic amino acids (Lys, Arg) and cysteine (Cys)
3. Imino acid (Pro) and glycine
4. Acidic amino acids (Asp, Glu)
5. Dipeptides and tripeptides.
   Moreover, glutathione also helps in absorption.

Table 14.5. Abnormalities in absorption

1. The deficiency of the enzyme 5-oxoprolinase leads to **oxoprolinuria** (pyroglutamic aciduria).
2. The allergy to certain food proteins (milk, fish) is due to absorption of partially digested proteins.
3. Defects in the intestinal amino acid transport systems are seen in inborn errors of metabolism.
   3-A. Hartnup's disease
   3-B. Imino glycinuria
   3-C. Cystinuria (see Chapter 16)
   3-D. Lysinuric protein intolerance

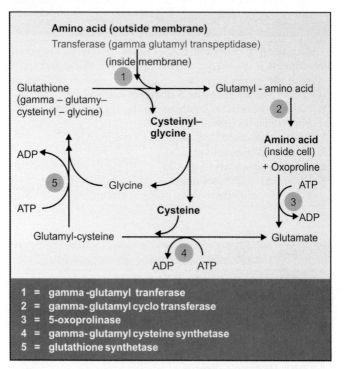

1 = gamma-glutamyl tranferase
2 = gamma-glutamyl cyclo transferase
3 = 5-oxoprolinase
4 = gamma-glutamyl cysteine synthetase
5 = glutathione synthetase

Fig. 14.4. Gamma glutamyl cycle (Miester cycle) by which amino acid is absorbed into the cell

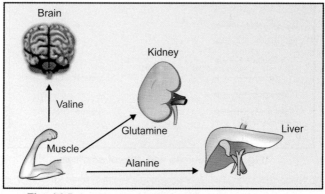

Fig. 14.5. Inter-organ transport of amino acids during fasting conditions

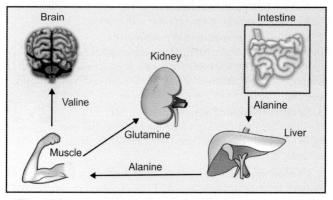

Fig. 14.6. Inter-organ transport of amino acids after taking food (post-prandial condition)

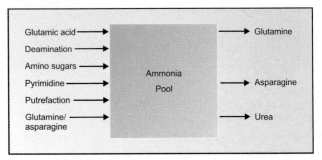

Fig. 14.7. Sources and fate of ammonia

Table 14.6. Significance of transamination

1. First step of catabolism
In this first step, **ammonia** is removed, and then rest of the amino acid is entering into catabolic pathway.

2. Synthesis of non-essential amino acids
By means of transamination, all non-essential amino acids could be synthesised by the body from keto acids available for other sources, e.g. **pyruvate** could be transaminated to synthesise **alanine**.

3. Interconversion of amino acids
If amino acid no.1 is high and no.2 is low; the amino group from no. 1 could be transferred to alpha keto acid to give amino acid no. 2.

4. Clinical significance of transamination
Aspartate amino transferase (**AST**) is increased in **myocardial infarction** and alanine amino transferase (**ALT**) in **liver** diseases.

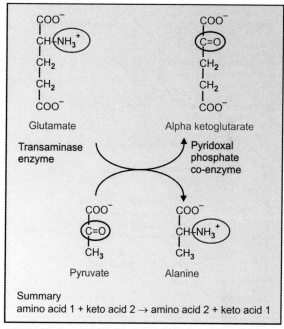

Glutamate

Alpha ketoglutarate

Transaminase enzyme

Pyridoxal phosphate co-enzyme

Pyruvate

Alanine

Summary

amino acid 1 + keto acid 2 → amino acid 2 + keto acid 1

Fig. 14.8. Transamination reaction. In this example, enzyme is alanine amino transferase (ALT) and pyridoxal phosphate is the co-enzyme. The reaction is readily reversible

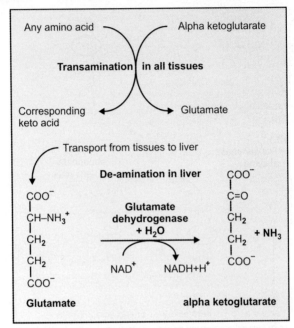

Fig. 14.9. Transamination + deamination = transdeamination. Only liver contains glutamate dehydrogenase which deaminates glutamate. So, all amino acids are first transaminated to glutamate, which is then finally deaminated (transdeamination)

Table 14.7. Disposal of ammonia

1. First line of Defense (Trapping of ammonia)
   The intracellular ammonia is immediately trapped by glutamic acid to form glutamine, especially in brain cells.
2. Transportation of Ammonia
   Glutamic acid is the major transport form of ammonia from the tissues to the liver. **Glutamine** and asparagine are the transport forms of ammonia from brain.
3. Final disposal
   The ammonia is **detoxified to urea by liver** cells, and then excreted through kidneys. **Urea is the end product of protein metabolism.**

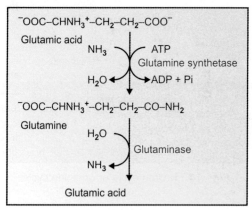

Fig.14.10. Ammonia trapping as glutamine

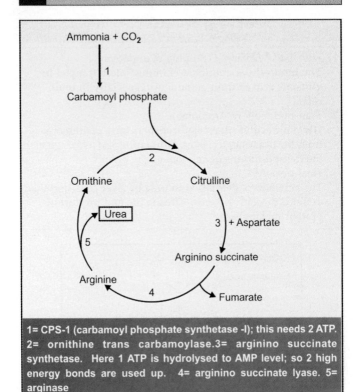

1= CPS-1 (carbamoyl phosphate synthetase -I); this needs 2 ATP. 2= ornithine trans carbamoylase. 3= arginino succinate synthetase. Here 1 ATP is hydrolysed to AMP level; so 2 high energy bonds are used up. 4= arginino succinate lyase. 5= arginase

Fig. 14.11. Urea cycle or ornithine cycle

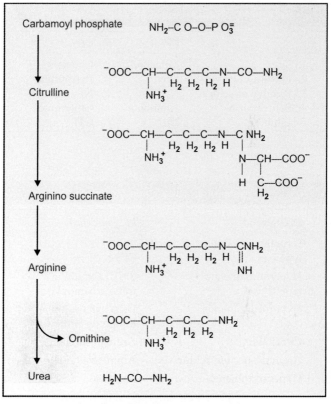

Fig. 14.12. Components of urea cycle

Table 14.8. Comparison: CPS I and II enzymes

	CPS-I	CPS-II
1. Site	Mitochon	Cytosol
2. Pathway of	Urea	Pyrimidine
3. Positive effector	NAG	Nil
4. Source for N	Ammonia	Glutamine
5. Inhibitor	Nil	CTP

Table 14.9. Urea cycle disorders

Diseases	Enzyme deficit
Hyperammonemia type I	CPS-I
Hyperammonemia type II	(OTC) Ornithine transcarbamoylase
Hyperornithinemia	Defective ornithine tranporter protein
Citrullinemia	Argininosuccinate
Argininosuccinic aciduria	Argininosuccinate lyase
Hyperargininemia	Arginase

Table 14.10. Urea level in blood

In clinical practice, blood urea level is taken as an **indicator of renal function**. The normal urea level in plasma is from **20 to 40 mg/dl** (2.4-4.8 mmol/L). (Urea Nitrogen = 8-20 mg/dl or 3-9 mmol/L). Blood urea level is increased where renal function is inadequate. Urinary excretion of urea is 15 to 30 g/day (6-15 g nitrogen/day). Urea constitutes 80% of urinary organic solids.

Table 14.11. One-carbon compounds; THFA = tetrahydrofolic acid

Group	Structure	Carried by
Formyl	–CHO	N^5–formyl–THFA and N^{10}–formyl–THFA
Formimino	–CH=NH	N^5–formimino–THFA
Methenyl	=CH–	N^5,N^{10}–methenyl–THFA
Hydroxymethyl	–CH$_2$OH	N^{10}–hydroxymethyl–THFA
Methylene	–CH$_2$–	N^5,N^{10}–methylene–THFA
Methyl	–CH$_3$	N^5–methyl–THFA and methyl cobalamin

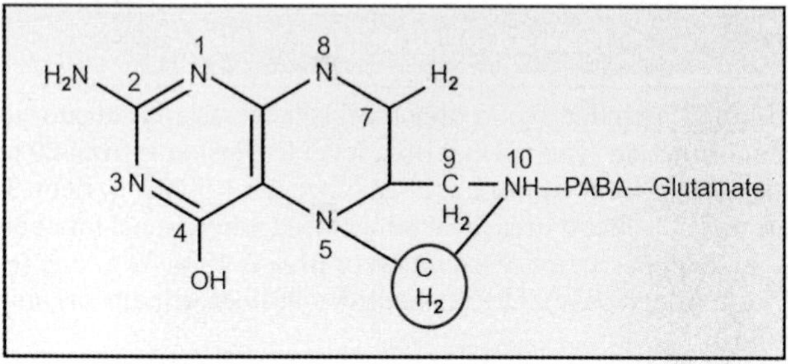

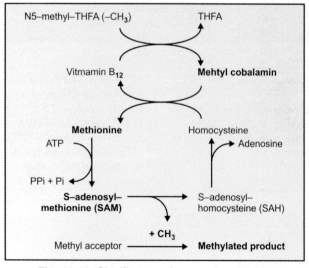

Fig. 14.13. Tetrahydrofolic acid (THFA). Methylene group is attached to N-5 and N-10

Fig. 14.14. Significance of one-carbon groups

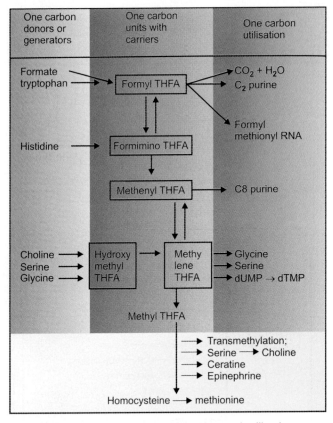

Fig. 14.15. One-carbon generation and utilisation

15. Glycine Metabolism

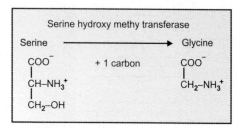

Fig.15.1. Formation of glycine from serine

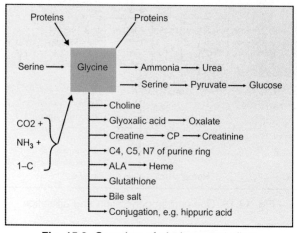

Fig. 15.2. Overview of glycine metabolism

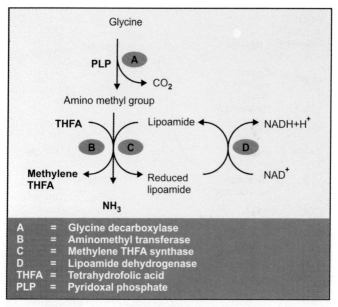

Fig. 15.3. Glycine cleavage system. Glycine is completely degraded to CO_2, ammonia and one-carbon unit methylene THFA. The reactions are readily reversible, when the enzymes are together called glycine synthase

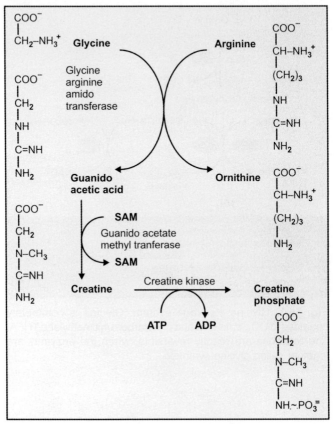

Fig. 15.4. Creatine metabolism

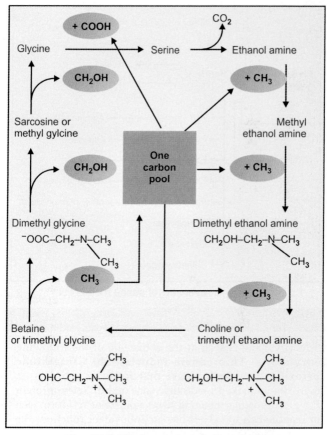

Fig. 15.5. Glycine-serine-choline cycle

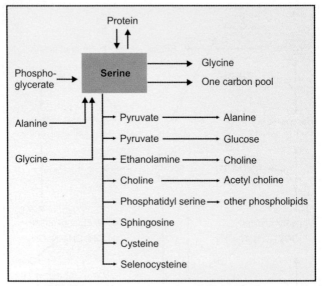

Fig. 15.6. An overview of serine metabolism

Table 15.1. Selenocysteine

Selenocysteine is seen at the active site of the following enzymes: **a) Thioredoxin reductase; b) Glutathione peroxidase; c) De-iodinase** that removes iodine from thyroxine to make tri-iodo-thyronine and **d) Selenoprotein P**, a glycoprotein containing 10 selenocysteine residues, seen in mammalian blood. It has an antioxidant function. Its concentration falls in selenium deficiency.

16. Cysteine and Methionine

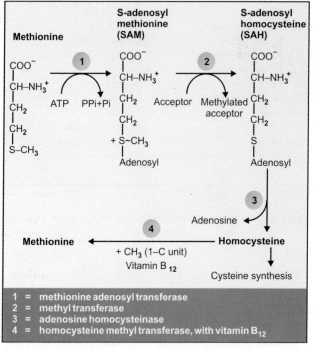

Fig. 16.1. Formation of active methionine

Table 16.1. Transmethylation reactions

Methyl acceptors	Methylated products
Guanido acetic acid	Creatine
Nicotinamide	N-methyl nicotinamide
Norepinephrine	Epinephrine
Epinephrine	Metanephrine
Norepinephrine	Normetanephrine
Ethanolamine	Choline
Carnosine	Anserine
Acetyl serotonin	Melatonin
Serine	Choline
Histidine	Methyl histidine
tRNA	Methylated tRNA

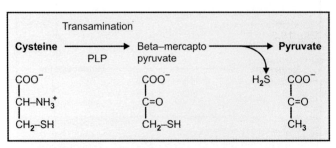

Fig. 16.2. Pyruvate formation from cysteine

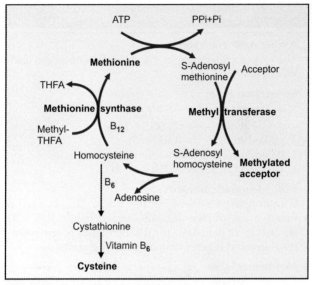

Fig. 16.3. Summary of methionine to cysteine conversion.
Note the role played by vitamins

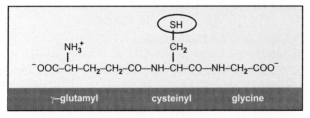

Fig. 16.4. Glutathione

Table 16.2. Functions of glutathione

1. **Amino Acid Transport**
   The role of glutathione in the absorption of amino acid is shown in Figure 14.4.

2. **Co-enzyme Role**
   Metabolic role of GSH is mainly in reduction reactions

   $$2GSH \longrightarrow GS\text{-}SG + H_2$$
   (Reduced)   (Oxidised)

   i. Maleayl aceto acetate → fumaryl acetoacetate
   ii. (Iodine) I_2 + 2GSH → 2HI + GS-SG

3. **RBC Membrane Integrity**
   Glutathione is present in the RBCs. This is used for inactivation of free radicals formed inside RBC. The occurrence of hemolysis in GPD deficiency is attributed to the decreased regeneration of reduced glutathione.

4. **Met-hemoglobin**
   The met-Hb is unavailable for oxygen transport. Glutathione is necessary for the reduction of met-hemoglobin (ferric state) to normal Hb (ferrous state).

   $$2Met\text{-}Hb\text{-}(Fe^{3+}) + 2GSH \rightarrow$$
   $$2Hb\text{-}(Fe^{2+}) + 2H^+ + GS\text{-}SG$$

5. **Conjugation for Detoxification**
   Glutathione helps to detoxify **a)** organo phosphorus compounds; **b)** halogenated compounds; **c)** heavy metals; **d)** drug metabolism. The reaction is catalysed by **glutathione-S-transferase** (GST)

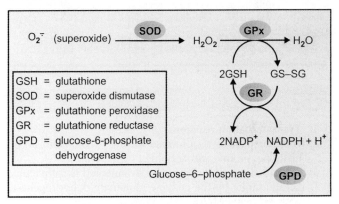

Fig. 16.5. Free radical scavenging

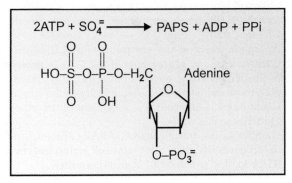

Fig. 16.6. Phosphoadenosine phosphosulphate (PAPS)

17. Acidic, Basic and Branched Chain Amino Acids

Table 17.1. Functions of glutamic acid

1. **Transamination reactions:** (Fig. 14.8).
2. Glutamic acid is also formed during the metabolism of histidine, proline and arginine.
3. **Deamination:** Glutamic acid is deaminated to form alpha ketoglutarate by the enzyme glutamate dehydrogenase with the help of NAD^+ (Fig. 14.9).
4. **Glucogenic:** Glutamic acid enters the TCA cycle, becomes oxaloacetate and goes to **glucogenic** pathway.
5. **N-acetyl glutamate** (NAG) is a positive modifier of carbamoyl phosphate synthetase-I in the mitochondria. Glutamic acid + Acetyl CoA → NAG + CoASH
6. **Ammonia trapping and glutamine** (Fig.14.10):
7. **Gamma carboxy glutamic acid** (GCGA) is present in prothrombin. The gamma carboxyl group is added as a post-translational modification, which needs vitamin K.
9. **Glutathione:** Glutamate is a constituent of the tripeptide glutathione (Fig. 16.4).
10. **Gamma amino butyric acid (GABA):** Glutamic acid on decarboxylation gives rise to **gamma amino butyric acid (GABA)**. It is an inhibitory neurotransmitter.

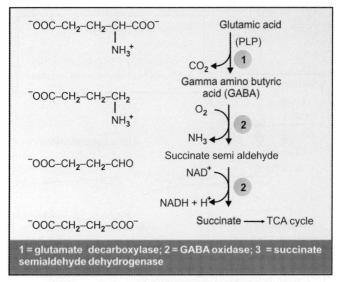

Fig. 17.1. GABA metabolism

GABA is an inhibitory neurotransmitter, because it opens the **chloride channels** in post-synaptic membranes in CNS. Both the formation and catabolism of GABA require pyridoxal phosphate (PLP) as co-enzyme (Steps 1 and 2). Therefore in pyridoxine deficiency, this pathway is affected, leading to **convulsions**. Sodium **valproate** which inhibits GABA oxidase is used in the treatment of epilepsy.

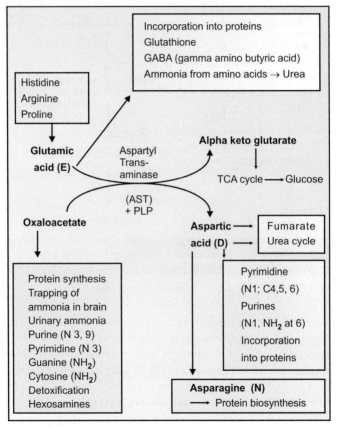

Fig. 17.2. Metabolisms of acidic amino acids

Table 17.2. Branched chain amino acids

1. **Valine** (Val) (V)is glucogenic; **Leucine** (Leu) (L) is ketogenic while **Isoleucine** (Ile) (I) is both ketogenic and glucogenic. All the three are **essential** amino acids. They serve an important role as an alternate source of **fuel for the brain** especially under conditions of starvation.

2. **Maple Syrup Urine Disease (MSUD)**
   i. It is also called branched chain ketonuria. The incidence is 1 per 1 lakh births. The name originates from the characteristic smell of urine (similar to burnt sugar or maple sugar) due to excretion of branched chain keto acids.
   ii. The basic biochemical defect is **deficient decarboxylation** of branched chain keto acids (BKA).
   iii. Clinical findings: Convulsions, severe **mental retardation**, acidosis, coma and death within the first year of life.
   iv. Urine contains **branched chain keto acids** and amino acids. Rothera's test is positive.
   v. Treatment: Giving a diet low in branched chain amino acids. Mild variants will respond to high doses of thiamine. This is because the decarboxylation of the BKA requires **thiamine**

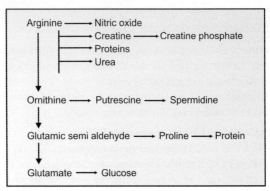

Fig.17.3. Metabolism of arginine and ornithine

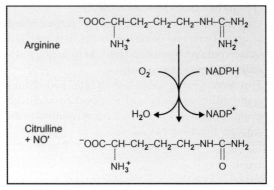

Fig.17.4. Nitric oxide synthase (NOS) reaction. The enzyme contains heme, FAD, FMN, tetrahydro-biopterine and calmodulin. It utilises NADPH

Table 17.3. Nitric oxide

Nitric oxide is an uncharged molecule having an unpaired electron, so it is a highly reactive "free radical". It is correctly written with a superscript dot (NO'). Nitric oxide is formed from arginine by the enzyme **nitric oxide synthase (NOS)** (Fig. 17.4).

Mechanism of Action of Nitric Oxide: NO diffuses to the adjacent smooth muscle and activates **guanylate cyclase**. Increased level of **cyclic GMP** activates protein kinase in smooth muscles, which causes dephosphorylation of myosin light chains, leading to relaxation of muscles.

Physiological Actions of Nitric Oxide:
1. **Blood vessels:** NO is a potent **vasodilator**. The normal blood pressure is maintained by the NO liberated by endothelial NOS.
2. **Platelets:** NO inhibits adhesion of platelets and so depresses platelet functions.
3. **Central nervous system:** In CNS, glutamate acts on N-methyl-D-aspartate (NMDA) receptors to cause a long-standing calcium influx. This activates NOS. NO stimulates the releasing hormones (CRH, GHRH and LHRH).
4. **Macrophages:** They produce NO' ; this is lethal to **micro-organisms**. This is induced by interleukin and tumor necrosis factor.

Table 17.4. Nitric oxide in diseases and treatment

1. **Angina Pectoris:** Nitroprusside can directly release NO. Nitroglycerine (glyceryl trinitrite) requires glutathione to produce NO. These will dilate coronary arteries; and are beneficial in treating angina pectoris.

2. **Pulmonary Hypertension:** Inhalation of NO is useful in the treatment of pulmonary hypertension and high altitude pulmonary edema. NO produces pulmonary vasodilatation, without lowering systemic blood pressure.

3. **Impotence:** NO relaxes smooth muscles in the corpus cavernosum and increases blood flow into the penis and makes it erect. **Sildenafil** citrate (Viagra) selectively inhibits the specific phospho-diesterase type 5 (PDE-5); thus inhibiting hydrolysis of cGMP; and increasing the concentration of cGMP in corpus cavernosum.

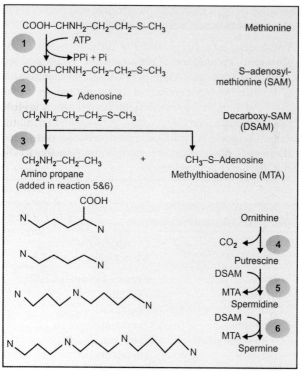

Fig. 17.5. Polyamine synthesis; 1 = methionine adenosyl transferase; 2 = SAM decarboxylase; 3= reaction coupled with 5 and 6; 4 = Ornithine decarboxylase; 5 = Spermidine synthase; 6 = Spermine synthase. SAM = S-adenosyl methionine; DSAM = decarboxy-SAM; MTA = methyl thio-adenosine

Table 17.5. Biochemical functions of polyamines

Several roles are suggested for polyamines, e.g. cell proliferation, stabilisation of ribosomes and DNA, synthesis of DNA and RNA, protection of DNA against depurination, etc. Polyamine concentration is increased in **cancer** tissues. Polyamines are **growth factors** in cell culture systems. Methyl thio adenosine (MTA) (step 3) has growth inhibitory activity. Decarboxylation of lysine produces cadaverine, which is also seen in mammalian tissues. It is also sometimes included in the list of polyamines.

Table 17.6. Biogenic amines

Substrates	Decarboxylated products, amine
Serine	Ethanol amine → Choline
Tyrosine	Tyramine
DOPA	Dopamine
Tryptophan	Tryptamine
5-OH-tryptophan	Serotonin
Histidine	Histamine
Ornithine	Putrescine
Lysine	Cadaverine
Cysteine	Taurine

18. Metabolism of Aromatic Amino Acids

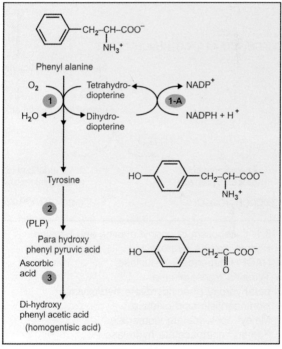

Fig. 18.1.

Contd...

Contd...

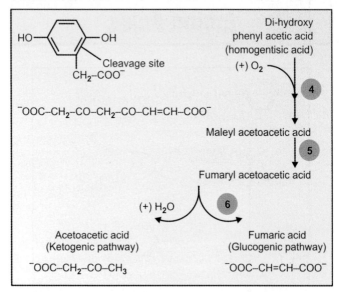

Fig. 18.1. Catabolism of phenyl alanine and tyrosine

1 = Phenyl alanine hydroxylase
1A = NADPH dependent reductase
2 = Tyrosine transaminase
3 = parahydroxy phenylpyruvate hydroxylase
4 = Homogentisic acid oxidase
5 = Maleyl acetoacetate isomerase
6 = Fumaryl acetoacetate hydrolase

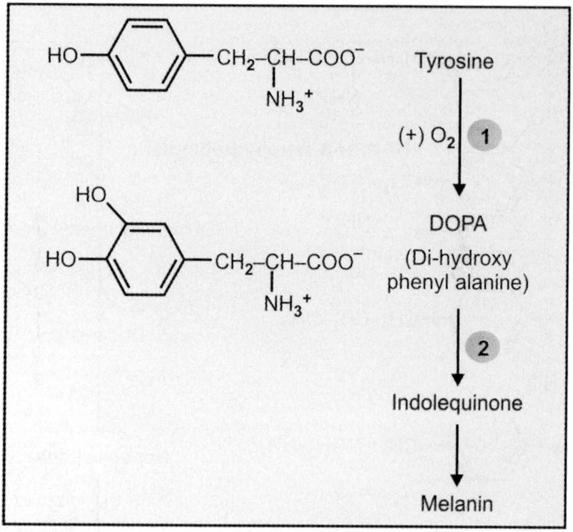

Fig. 18.2. Melanin synthesis pathway; 1 and 2 steps have the same enzyme, tyrosinase

Table 18.1. Tyrosinase and tyrosine hydroxylase

Both these enzymes will add hydroxyl group to tyrosine to produce dihydroxy phenyl alanine (**DOPA**). **Tyrosinase** is present in melanoblasts. It is used for **melanin** synthesis. But **Tyrosine hydroxylase** is present in adrenal medulla and it is used for **epinephrine** synthesis. Thus even in Tyrosinase-deficient person (albinism), epinephrine synthesis is normal

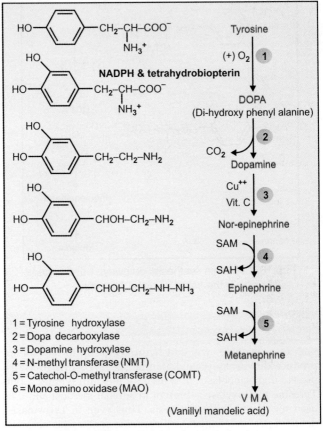

Fig. 18.3. Metabolism of catecholamines

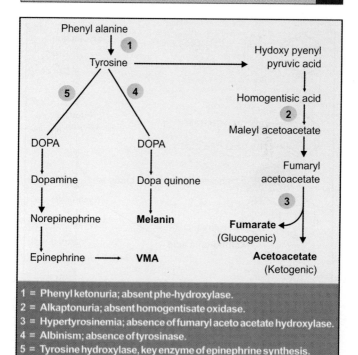

1 = Phenyl ketonuria; absent phe-hydroxylase.
2 = Alkaptonuria; absent homogentisate oxidase.
3 = Hypertyrosinemia; absence of fumaryl aceto acetate hydroxylase.
4 = Albinism; absence of tyrosinase.
5 = Tyrosine hydroxylase, key enzyme of epinephrine synthesis.

Fig. 18.4. Summary of tyrosine metabolism

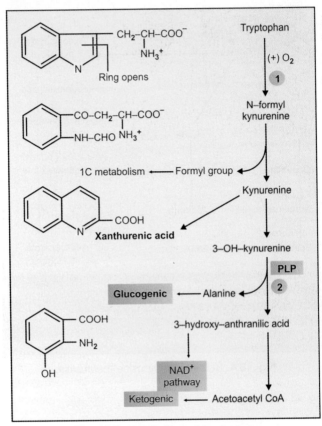

Fig. 18.5. Metabolism of tryptophan.
1= Tryptophan pyrrolase. 2 = Kynureninase

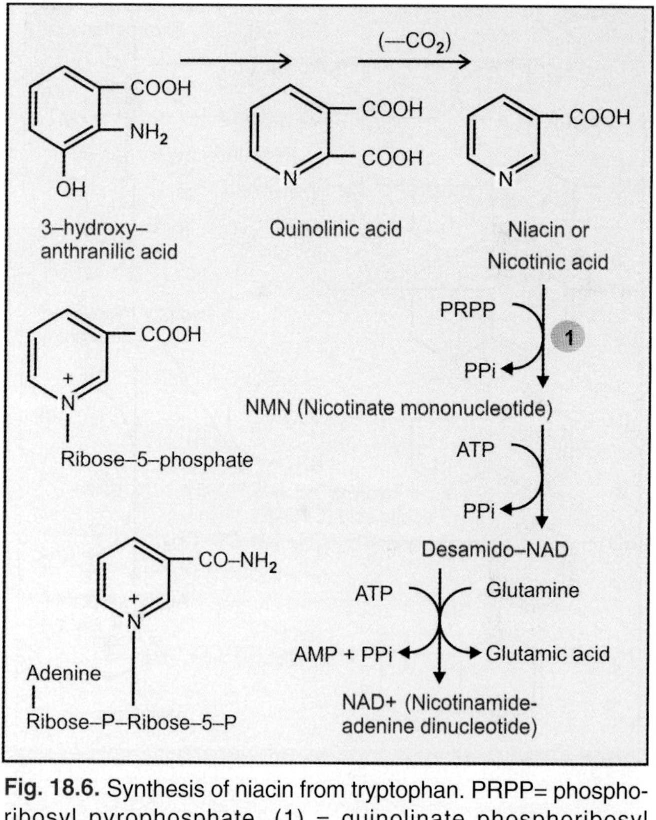

Fig. 18.6. Synthesis of niacin from tryptophan. PRPP= phosphoribosyl pyrophosphate. (1) = quinolinate phosphoribosyl transferase. About 60 mg of tryptophan will be equivalent to 1 mg of nicotinic acid

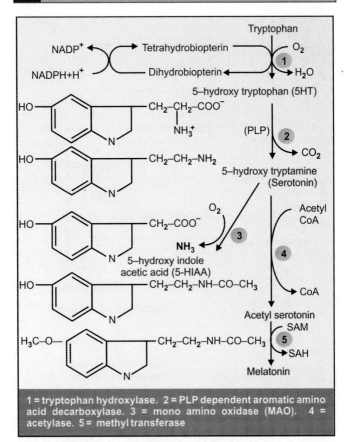

Fig. 18.7. Serotonin and melatonin synthesis

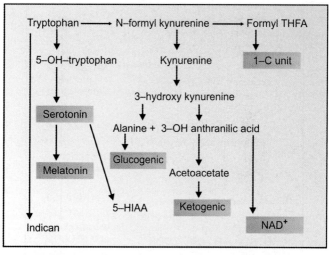

Fig. 18.8. Summary of tryptophan metabolism

Table 18.2. Melanin and melatonin are different

Melanin is the pigment of hair and skin; it is synthesised from tyrosine (Fig. 18.2)

Melatonin is a neurotransmitter synthesised from tryptophan (Fig. 18.7).

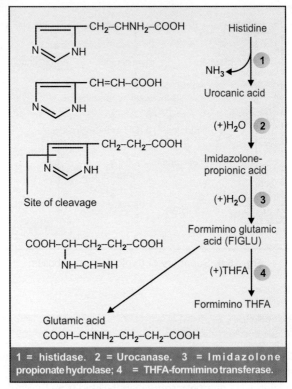

Fig. 18.9. Metabolism of histidine

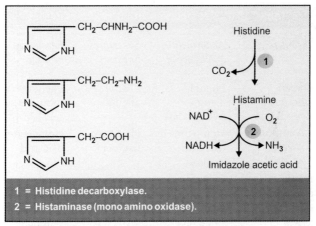

1 = Histidine decarboxylase.
2 = Histaminase (mono amino oxidase).

Fig. 18.10. Metabolism of histamine

Table 18.3. Summary of action of histamine

Tissues	Effects
1. Blood vessels	Pulmonary venous dilation; Large veins, smaller venules and capillaries are dialated
2. Cardiovascular system	Fall in BP; increased capillary permeability
3. Smooth muscles	Direct stimulant; contraction of bronchial muscles; bronchospasm
5. Exocrine glands	Stimulates gastric acid secretion

Table 18.4. Selected important amino acidurias

Disorders	Abnormalities or absence of	Substances in blood	Substances in urine
Phenyl ketonuria (type I)	Phenyl alanine hydroxylase	Phenyl alanine	Phenyl pyruvate
Alkaptonuria	Homogentisic acid oxidase	Homogentisic acid	Homogentisic acid
Homocystinuria (type 1)	Cystathionine beta synthase	Homocysteine; Methionine	Homocystine
Maple syrup urine disease	Branched chain keto acid decarboxylase	Val; Leu; Ile; keto acids	Val; Leu; Ile; keto acids
Methyl malonic aciduria	Methyl malonyl CoA mutase	Methyl malonic acid	Methyl maIonic acid; ketone bodies
Cystathioninuria	Cystathionase	Cystathionine	Cystathionine
Citrullinemia	Argininosuccinate synthetase	Ammonia; citrulline	Citrulline
Argininemia	Arginase	Arginine; ammonia	Arginine, ornithine

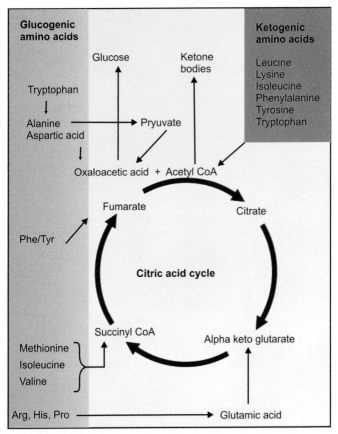

Fig. 18.11. Summary of metabolic fates of amino acids

Section 5

Integration of Metabolism

19. Citric Acid Cycle

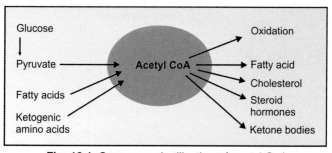

Fig. 19.1. Sources and utilisation of acetyl CoA

Table 19.1. Functions of the citric acid cycle

1. It is the final common oxidative pathway that oxidises acetyl CoA to CO_2.
2. It is the source of reduced co-enzymes that provide the substrate for the respiratory chain.
3. It acts as a link between catabolic and anabolic pathways (amphibolic role).
4. It provides precursors for synthesis of amino acids and nucleotides.
5. Components of the cycle have a direct or indirect controlling effects on key enzymes of other pathways.

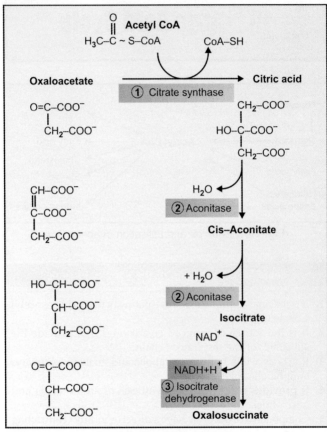

Fig. 19.2.

Contd...

Contd...

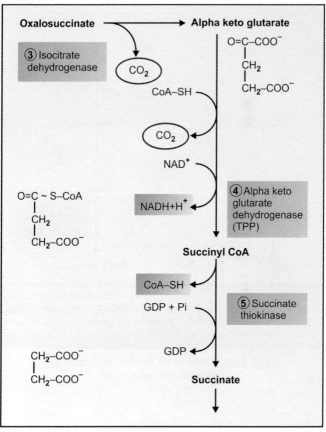

Fig. 19.2.

Contd...

Contd...

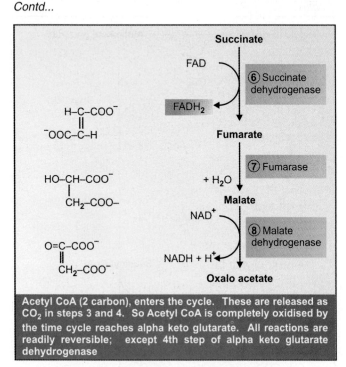

Acetyl CoA (2 carbon), enters the cycle. These are released as CO_2 in steps 3 and 4. So Acetyl CoA is completely oxidised by the time cycle reaches alpha keto glutarate. All reactions are readily reversible; except 4th step of alpha keto glutarate dehydrogenase

Fig. 19.2. Krebs cycle or citric acid cycle or tricarboxylic acid cycle

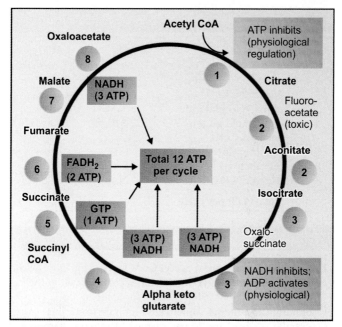

Fig. 19.3. Summary of Krebs citric acid cycle. Enzymes are numbered and marked in green rounds. Steps where energy is trapped are marked in red color. A total of 12 ATPs are generated during one cycle. Reactions numbers 3 and 4 are carbon dioxide elimination steps. Physiological regulatory steps are shown as blue backgrounds. Step no. 1 (citrate synthase) is physiologically inhibited by ATP. Step no. 3 (ICDH) is inhibited by NADH and activated by ADP

Table 19.2. ATP generation in citric acid cycle

Step no	Reactions	Co-enzyme	ATP generated
3	Isocitrate→ alpha keto glutarate	NADH	3
4	Alpha keto glutarate → succinyl CoA	NADH	3
5	Succinyl CoA → Succinate	GTP	1
6	Succinate → Fumarate	FADH$_2$	2
8	Malate → Oxaloacetate	NADH	3
	Total ATP per cycle		**12**

Table 19.3. Significance of TCA cycle

1. **Complete oxidation of acetyl CoA.** During the citric acid cycle, two carbon dioxide molecules are removed in the following reactions: **Step 3**, oxalosuccinate to alpha keto glutarate and **Step 4**, alpha keto glutarate to succinyl CoA. Acetyl CoA contains 2 carbon atoms. These two carbon atoms are now removed as CO_2 in steps 3 and 4. Net result is that **acetyl CoA is completely oxidised during one turn of cycle.**
2. **ATP generating steps in TCA cycle.** See Fig. 19.3 and Table 19.2.
3. **Final common oxidative pathway.** Citric acid cycle is the final common oxidative pathway of all foodstuffs. All the

Contd...

Contd...

major ingredients of foodstuffs are finally oxidised through the TCA cycle. **Carbohydrates** are metabolised through glycolytic pathway to pyruvate, then to acetyl CoA, which enters the citric acid cycle. **Fatty acids** through beta oxidation, are broken down to acetyl CoA and then enters this cycle. **Amino acids** after transamination enter into some or other points in this cycle.

4. **Fat is burned on the wick of carbohydrates.** Oxidation of fat (acetyl CoA) needs the help of oxaloacetate. The major source of oxaloacetate is pyruvate (carbohydrate). **Fats are burned under the fire of carbohydrates.**

5. **Excess carbohydrates are converted as neutral fat.** The pathway is glucose to pyruvate to acetyl CoA to fatty acid. However, **fat cannot be converted to glucose** because pyruvate dehydrogenase reaction (pyruvate to acetyl CoA) is an **absolutely irreversible** step.

6. **Amphibolic pathway** (catabolic + anabolic). There is a continuous influx (pouring into) and a continuous efflux (removal) of 4-carbon units from the TCA cycle (Figs 19.4 and 19.5).

7. **Anaplerotic role.** Anaplerotic reactions are "**filling up**" reactions or "influx" reactions which supply 4-carbon units to the TCA cycle (Fig. 19.5).

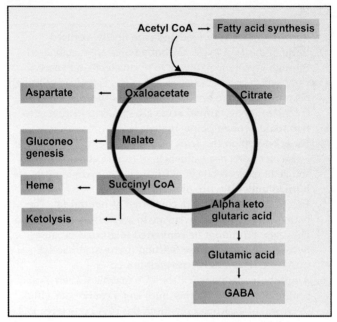

Fig. 19.4. Efflux of TCA cycle intermediates. The TCA cycle memebrs are shown with red background. Synthesized products are shown in green boxes

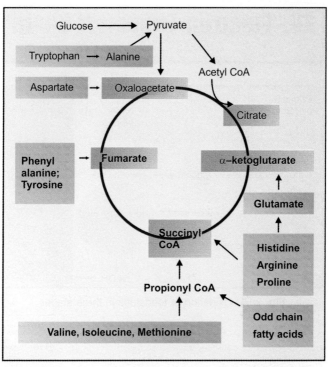

Fig. 19.5. Influx of TCA cycle intermediates. The TCA cycle memebrs are shown with red background. The precursors are shown in blue boxes

20. Electron Transport Chain

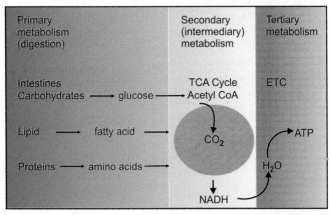

Fig. 20.1. Oxidation of foodstuffs in three stages

Table 20.1. Redox

Oxidation is defined as the loss of electrons and reduction as the gain in electrons. When a substance exists both in the reduced state and in the oxidised state, the pair is called a **redox couple**.

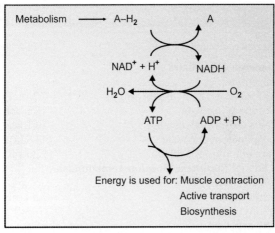

Fig. 20.2. Food is catabolised; energy from food is trapped as ATP; it is then used for body anabolism

Table 20.2. Electron transport chain

The electron flow occurs through successive dehydrogenase enzymes, known as electron transport chain (ETC). **The electrons are transferred from higher potential to lower potential.** The free energy change between NAD^+ and water is equal to 53 kcal/mol. This is so great that, if this much energy is released at one stretch, body cannot utilise it. Hence with the help of ETC assembly, the total energy change is released in small increments so that energy can be trapped as **chemical bond energy**.

Table 20.3. High-energy compounds

Energy rich compounds	$G^{0'}$ in kcal/mol
High energy phosphates	
1. Nucleotides:	
(ATP, GTP, UTP, UDP-glucose)	– 7.3 kcal
2. Creatine phosphate	– 10.5 kcal
3. 1,3-bisphospho glycerate	– 10.1 kcal
4. Phospho enol pyruvate	– 14.8 kcal
5. Inorganic pyrophosphate	– 7.3 kcal
High energy thioesters (Sulphur compounds)	
6. CoA derivatives:	
Acetyl CoA	– 7.5 kcal
Succinyl CoA	
Fatty acyl CoA	
HMG CoA	
7. S-adenosyl methionine (SAM)	– 7.0 kcal

Table 20.4. Summary of electron flow in ETC

Complex I: NADH → FMN → Fe-S → Co Q →
Complex II: Succinate → FAD → Fe-S → Co Q →
Complex III: Co Q → Fe-S → cyt.b → cyt.c1 → cyt. c
Complex IV: Cyt. c → cyt a-a3 → O_2

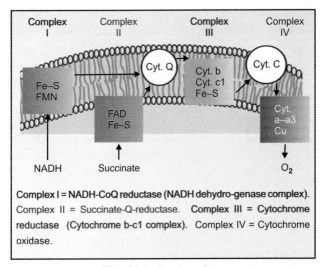

Complex I	Complex II	Complex III	Complex IV

Cyt. Q

Cyt. C

Fe–S
FMN

Cyt. b
Cyt. c1
Fe–S

FAD
Fe–S

Cyt.
a–a3
Cu

NADH   Succinate   O$_2$

Complex I = NADH-CoQ reductase (NADH dehydro-genase complex). Complex II = Succinate-Q-reductase.   Complex III = Cytochrome reductase (Cytochrome b-c1 complex). Complex IV = Cytochrome oxidase.

Fig. 20.3. Electron flow

Table 20.5. Organization of ET chain

The electrons are transferred from NADH to a chain of electron carriers. The electrons flow from the more electronegative components to the more electropositive components. All the components of electron transport chain (ETC) are located in the **inner membrane of mitochondria**. There are four distinct multi-protein complexes; these are named as **complex-I, II, III and IV.** These are connected by two mobile carriers, **co-enzyme Q** and **cytochrome c.**

Table 20.6. Substrates for dehydrogenases

NAD⁺ linked dehydrogenases

NAD$^+$ is derived from nicotinic acid, a member of the vitamin B complex. When the NAD$^+$ accepts the two hydrogen atoms, one of the hydrogen atoms is removed from the substrate as such. The other hydrogen atom is split into one hydrogen ion and one electron. The electron is also accepted by the NAD$^+$ so as to neutralize the charge on the co-enzyme molecule. The remaining hydrogen ion is released into the surrounding medium.

$$H_2 \longrightarrow H + H^+ + e^-$$
$$AH_2 + NAD^+ \rightarrow A + NADH + H^+$$

The NAD$^+$ linked dehydrogenases are:

  i. Glyceraldehyde-3-phosphate dehydrogenase
 ii. Isocitrate dehydrogenase
iii. Glutamate dehydrogenase
 iv. Beta hydroxy acyl CoA dehydrogenase
  v. Pyruvate dehydrogenase
 vi. Alpha keto glutarate dehydrogenase

FAD-linked dehydrogenases

When FAD is the co-enzyme, (unlike NAD$^+$), both the hydrogen atoms are attached to the flavin ring. Examples:

 i. Succinate dehydrogenase
ii. Fatty acyl CoA dehydrogenase

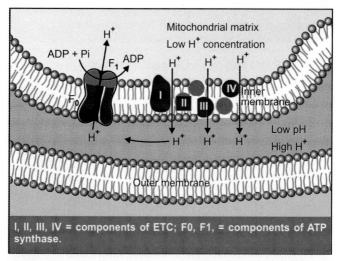

I, II, III, IV = components of ETC; F0, F1, = components of ATP synthase.

Fig. 20.4. Summary of ATP synthesis. ETC complexes will push hydrogen ions from matrix into the intermediate space. So, intermediate space has more H⁺ (highly acidic) than matrix. So, hydrogen ions tend to leak into matrix through Fo. Then ATPs are synthesized

Table 20.7. Compounds which affect electron transport chain and oxidative phosphorylation

1. **Site-1 (complex I to Co-Q) specific inhibitors**
   Barbiturates (amobarbital), sedative

2. **Site-2 (complex III to cytochrome c) inhibitors**
   BAL (British anti lewisite), antidote of war gas

3. **Site-3 (complex IV) inhibitors**
   Carbon monoxide, inhibits cellular respiration
   Cyanide (CN^-)

4. **Site between succinate dehydrogenase and Co-Q**
   Malonate, competitive inhibitor of succinate DH

5. **Inhibitors of oxidative phosphorylation**
   Atractyloside, inhibits translocase
   Oligomycin, inhibits flow of protons through Fo
   Ionophores, e.g., Valinomycin

6. **Uncouplers**
   2,4-dinitro phenol (2,4-DNP)

7. **Physiological uncouplers**
   Thyroxine, in high doses

Section 6

Proteins

21. Plasma Proteins

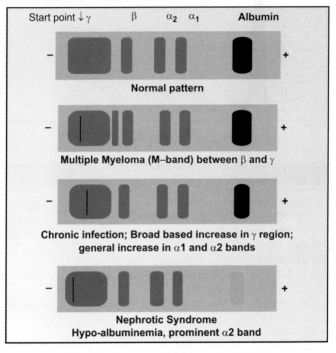

Fig. 21.1. Serum electrophoretic patterns

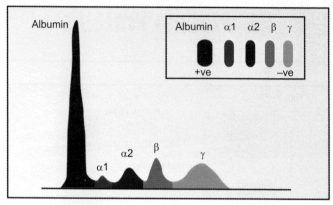

Fig. 21.2. Normal electrophoretic pattern

Table 21.1. Functions of albumin

1. Colloid osmotic pressure of plasma
   Proteins exert the **'effective osmotic pressure'**. It is about 25 mm Hg, and 80% of it is contributed by albumin. (See Fig. 21.3).
2. Transport function; Albumin is the carrier of:
   i. **Bilirubin** and **non-esterified fatty acids** are specifically transported by albumin.
   ii. **Drugs** (sulpha, aspirin, salicylates, dicoumarol, phenytoin).
   iii. **Hormones:** steroid hormones, thyroxine.
   iv. **Metals:** calcium, copper and heavy metals are non-specifically carried by albumin.

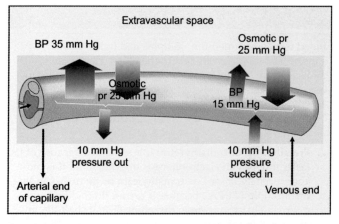

Fig. 21.3. Starling hypothesis. At the capillary end the blood pressure (BP) or hydrostatic pressure expels water out, and effective osmotic pressure (EOP) takes water into the vascular compartment. If protein concentration in serum is reduced, the EOP is correspondingly decreased. Then return of water into blood vessels is diminished, leading to accumulation of water in tissues. This is called edema. Total protein content of normal plasma is 6 to 8 g/100 ml. The plasma proteins consist of albumin (3.5 to 5 g/dl), globulins (2.5-3.5 g/dl) and fibrinogen (200-400 mg/dl). Almost all plasma proteins, except immunoglobulins are synthesized in liver

Table 21.2. Transport proteins

Blood is a watery medium; so lipids and lipid soluble substances will not easily mix in the blood. Hence such molecules are carried by specific carrier proteins. Important ones are described below:

1. **Albumin:** It is an important transport protein, which carries bilirubin, free fatty acids, calcium and drugs (see above).

2. **Pre-albumin or transthyretin:** It is so named because of its faster mobility in electrophoresis than albumin. It is more appropriately named as **transthyretin** or thyroxin binding pre-albumin (**TBPA**), because it carries thyroid hormones, thyroxin (T4) and tri-iodo thyronine (T3). Its half-life in plasma is only 1 day.

3. **Thyroxine binding globulin (TBG):** It is the specific carrier molecule for thyroxine and tri-iodo thyronine. TBG level is increased in pregnancy; but decreased in nephrotic syndrome.

4. **Retinol binding protein (RBP):** It carries vitamin A.

5. **Transcortin:** It is also known as **Cortisol binding globulin (CBG)**. It is the transport protein for cortisol and corticosterone.

6. **Transferrin:** It carries iron in plasma.

Table 21.3. Ceruloplasmin

  i. Ceruloplasmin is a plasma protein. It is blue in color. It is synthesized in liver. It contains 6 to 8 **copper atoms** per molecule.

 ii. Ceruloplasmin is also called **Ferroxidase**, an enzyme which helps in the incorporation of iron into transferrin.

iii. Copper is bound with albumin loosely, and so easily exchanged with tissues. Hence **transport protein for copper is albumin**.

 iv. Ceruloplasmin is an acute phase protein. So its level in blood may be increased in all inflammatory conditions, collagen disorders and in malignancies.

Wilson's disease

 a. Normal blood level of ceruloplasmin is 25-50 mg/dl. This level is reduced to less than 20 mg/dl in **Wilson's hepatolenticular degeneration.**

 b. It is an inherited autosomal recessive condition. Incidence of the disease is 1 in 50,000.

 c. Copper is not excreted through bile, and hence copper toxicity. So ceruloplasmin level in blood is decreased.

 d. Accumulation in liver leads to hepatocellular degeneration and **cirrhosis**. Deposits in brain basal ganglia leads to **lenticular degeneration** and neurological symptoms.

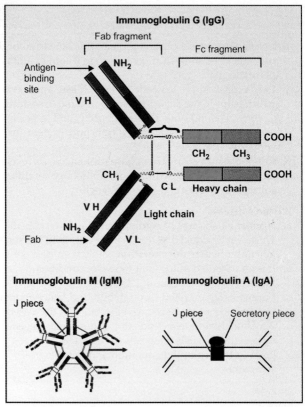

Fig. 21.4. Immunglobulin molecule. Red area is variabile region; VH= variable heavy region; VL= variable light chain; CH= constant heavy

Table 21.4. Different immunoglobulin classes

	IgG	IgA	IgM	IgE
Heavy chain	γ	α	μ	ε
No. of units (2L + 2H)	1	2	5	1
Additional unit	—	S and J	J piece	—
Mol wt (kD)	146	385	970	190
Sed. coefficient	7 S	11 S	19 S	8S
Carbohydrate	3(%)	10(%)	12(%)	11(%)
Normal serum level (mg/dl)	1000	200	100	0.003
Half-life in days	20	6	10	2

Table 21.5. Functions of immunoglobulins

	IgG	IgA	IgM	IgD	IgE
Placental transfer	+	–	–	–	–
Complement fixation	+	+	++	–	–
Agglutination	+	++	+++	–	–
Binding to macrophages	+	–	–	–	–
Fixation to mast cells	–	–	–	–	+
Primary antibody	–	–	+	–	–
External secretions	–	+	–	–	–
Natural antibodies	–	–	+	–	–

Table 21.6. Multiple myeloma (plasmacytoma)

When Ig-secreting cells are transformed into malignant cells, one clone alone is enormously proliferated. Thus, Ig molecules of the very same type are produced in large quantities.

This is seen in electrophoresis as the myeloma band or **monoclonal band** or M band with a sharp narrow spike (see Fig. 21.1).

There will be paraproteinemia, anemia, lytic bone lesions and proteinuria. Bone marrow examination reveals large number of malignant plasma cells.

Bone pain and tenderness are the common presenting complaints. Spontaneous pathological fracture of weight bearing bones, rib and vertebrae may occur.

Hypercalcemia and hypercalciuria are therefore common features.

About 70% cases have IgG and 25% have IgA paraproteins.

Total Ig content may be very high; but useful antibodies may be very low, so that general immunity is depressed and recurrent infections are common.

Raised beta-2-microglobulin (Mol. wt. 11,800) is another feature of multiple myeloma.

Table 21.7. Bence-Jones proteinuria

This disorder is seen in 20% of patients with multiple myeloma. Monoclonal light chains are excreted in urine. This is due to asynchronous production of H and L chains or due to deletion of portions of L chains, so that they cannot combine with H chains. The Bence-Jones proteins have the special property of precipitation when heated between 45°C and 60°C; but redissolving at higher than 80°C and lower than 45°C. Bradshaw's test is also positive, when a few ml of urine is layered over a few ml of concentrated hydrochloric acid, a white ring of precipitate is formed. These proteins may block kidney tubules, leading to renal failure. So, Bence-Jones myeloma has a poorer prognosis.

Table 21.8. Waldenstrom's macroglobulinemia

In this condition IgM level in blood is increased considerably with a monoclonal peak. This is due to malignant proliferation of IgM clones. Males are affected mostly. Since IgM are macromolecules, they may form aggregates or cryoprecipitates, serum viscosity is increased. Hyperviscous serum will form globular precipitate. Hyperviscosity leads to recurrent bleeding.

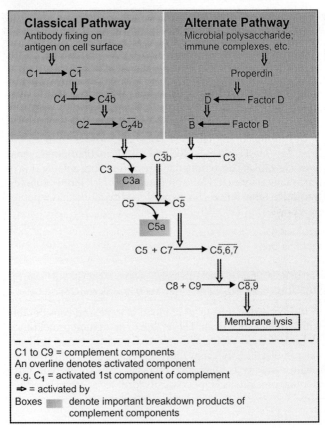

Fig. 21.5: Pathways of complement activation

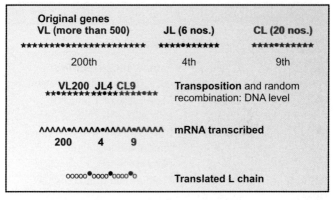

Fig. 21.6. Production of diversity in immunoglobulins. Somatic recombination in precursors of lymphocytes. Each L chain gene contains over 500 VL (variable light) segments, 5-6 JL (joining light) segments and 10-20 CL (constant light) segments. In this example, random rearrangement allows VL (200)-JL(4)-CL(9) segments to remain in the gene, while other regions of genes are deleted. This VL-JL-CL segments are transcribed as a single mRNA, and later translated into a specific immunoglobulin light chain. Another permutation is taking place in another cell. Thus millions of cells together can produce endlessly diverse light chains

Table 21.9. Selected important lymphokines

Name	Mol.wt.	Target cell / Function
IL-1	17,000	Induces IL-2 receptors; induces acute phase protein
IL-2	15,000	Maturation of T and NK cells into LAK cells
IL-6	20,000	B cell and myeloid differentiation; induces acute phase proteins
IFN-alpha		Proliferation of macros
IFN-beta		Antiviral
IFN-gamma	50,000	Antiviral; differentiates cells
TNF-alpha	17,000	Inflammation, fibrosis, pyrexia, acute phase proteins, necrosis
G-CSF	20,000	Stem cell stimulation of granulocytes
GM-CSF	22,000	Stem cell stimulation of granulocytes and macros
MIF	50,000	Activation and inhibition of mobility of macrophages

IL= Interleukin; IFN= Interferon; TNF= Tumor Necrosis
Factor; G-CSF= Granulocyte Colony Stimulating Factor;
GM-CSF = Granulocyte Macrophage Colony
Stimulating Factor; MIF= Macrophage Migration Inhibition Factor

22. Heme and Hemoglobin

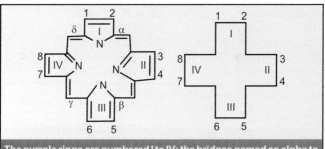

The pyrrole rings are numbered I to IV; the bridges named as alpha to delta and the possible sites of substitutions are denoted from 1 to 8.

Fig. 22.1: Porphyrin ring

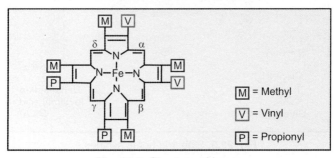

M = Methyl

V = Vinyl

P = Propionyl

Fig. 22.2: Structure of heme

Table 22.1. Porphyrins of biological importance. See also Fig. 22.2 for the structure of heme

Name of porphyrin	Order of substituents from 1st to 8th positions
Uroporphyrin I	A,P, A,P, A,P, A,P
Uroporphyrin III	A,P, A,P, A,P, P,A
Coproporphyrin I	M,P, M,P, M,P, M,P
Coproporphyrin III	M,P, M,P, M,P, P,M
Protoporphyrin III	M,V, M,V, M,P, P,M

(A = acetyl; P = propionyl; M = methyl; V = vinyl)

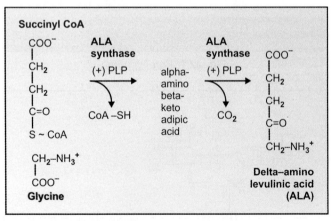

Fig. 22.3. Step 1 in heme synthesis

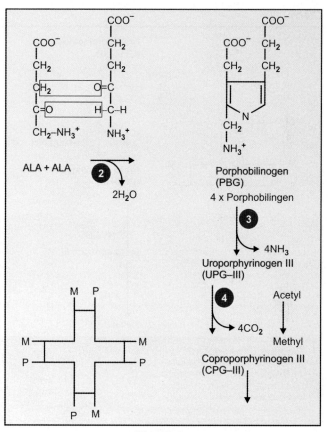

Fig. 22.4.

Contd...

Contd...

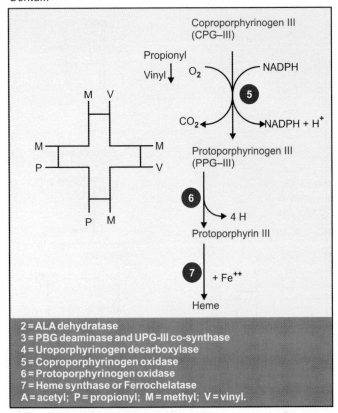

2 = ALA dehydratase
3 = PBG deaminase and UPG-III co-synthase
4 = Uroporphyrinogen decarboxylase
5 = Coproporphyrinogen oxidase
6 = Protoporphyrinogen oxidase
7 = Heme synthase or Ferrochelatase
A = acetyl; P = propionyl; M = methyl; V = vinyl.

Fig. 22.4. Steps of heme synthesis

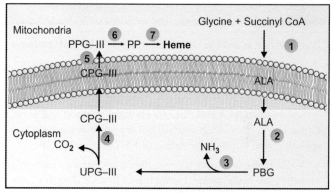

Fig. 22.5. Summary of heme biosynthesis. The numbers denote the enzymes. Part of synthesis is in mitochondria, and the rest in cytoplasm

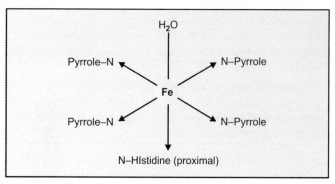

Fig. 22.6. In the heme molecule, iron atom is co-ordinately linked with nitrogen atoms

Table 22.2. Acute intermittent porphyria (AIP)

It is inherited as an autosomal **dominant** trait. **PBG-deaminase** is deficient.

This leads to a secondary increase in activity of ALA synthase, since the end-product inhibition is not effective.

The levels of **ALA and PBG are elevated** in blood and urine. Urine is colorless when voided, but the color is increased on standing due to photo-oxidation of PBG to porphobilin.

As the name indicates, the symptoms appear intermittently and they are quite vague. Hence it is at times called the "little imitator". Most commonly, patients present with **acute abdominal pain**. The patients often land up with the surgeon as a case of acute abdomen and on several instances exploratory laparotomies are done.

An attack is precipitated by starvation and symptoms are alleviated by a high carbohydrate diet. Drugs like barbiturates, which are known to induce ALA synthase, can precipitate an attack.

Diagnosis of Porphyrias: The presence of porphyrin precursor in urine is detected by Ehrlich's reagent. When urine is observed under ultraviolet light; porphyrins if present, will emit strong red fluorescence.

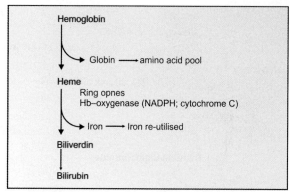

Fig. 22.7. Catabolic pathway of hemoglobin

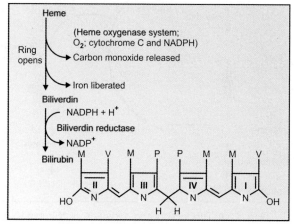

Fig. 22.8. Breakdown of heme

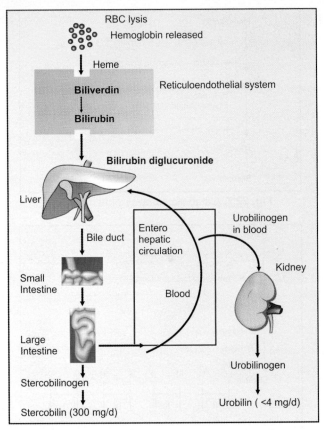

Fig. 22.9. Production and excretion of bilirubin

Table 22.3. Properties of conjugated and free bilirubin

	Free bilirubin	*Conjugated bilirubin*
In water	Insoluble	Soluble
In alcohol	Soluble	Soluble
Normal plasma level (mg/dl)	0.2-0.6	0-0.2
In bile	Absent	Present
In urine	Always absent	Normally absent
van den Berg's test	Indirect Positive	Direct Positive

Table 22.4. Differential diagnosis of jaundice

	Hemolytic jaundice	*Hepatocellular jaundice*	*Obstructive jaundice*
Blood			
Free bilirubin	Increase	Increase	Normal
Conj. bilirubin	Normal	Increase	Increase
ALP	Normal	Increase	Very high
Urine			
Bile salts	Nil	Nil	Present
Conj.bilirubin	Nil	Nil	Present
Bilinogens	Increase	Nil	Nil

Table 22.5. Hyperbilirubinemias

1. Congenital Hyperbilirubinemias

They result from abnormal uptake, conjugation or excretion of bilirubin due to inherited defects. **Crigler-Najjar Syndrome** is due to the defect in conjugation. In Type 1 (Congenital non-hemolytic jaundice), there is severe deficiency of UDP **glucuronyl transferase**. The children die before the age of 2. Unconjugated bilirubin level increases to more than 20 mg/dl, and kernicterus is resulted.

2. Acquired Hyperbilirubinemias

Physiological Jaundice is also called as neonatal hyperbilirubinemia. In all newborn infants mild jaundice appears. This transient hyperbilirubinemia is due to an accelerated rate of destruction of RBCs and also because of the immature hepatic system of conjugation of bilirubin. It disappears by the second week of life.

3. Hemolytic Jaundice

3-A. Hemolytic Disease of the Newborn results from incompatibility between maternal and fetal blood groups. **(Rh incompatibility).** When blood level is more than 20 mg/dl, the capacity of albumin to bind bilirubin is exceeded. In young children before the age of 1 year, the blood-brain barrier is not fully matured, and therefore free bilirubin enters the brain **(Kernicterus)**. It is deposited in brain, leading to mental retardation, and toxic encephalitis.

Contd...

Contd...

> **3-B. Hemolytic Diseases of Adults** are seen in increased rate of hemolysis. The characteristic features are increase in unconjugated bilirubin in blood, absence of bilirubinuria and excessive excretion of UBG in urine and SBG in feces (Table 22.4). Common causes are Congenital spherocytosis or Autoimmune hemolytic anemias.
>
> **4. Hepatocellular Jaundice**
> The most common cause is viral hepatitis, caused by Hepatitis viruses A, B, C, D or G. Conjugation in liver is decreased and hence free bilirubin is increased in circulation. (Table 22.4).
>
> **5. Obstructive Jaundice**
> Conjugated bilirubin is increased in blood, and it is excreted in urine. UBG will be decreased in urine or even absent (Fig. 22.9 and Table 22.4). Since no pigments are entering into the gut, the feces become clay colored. The common causes of obstructive jaundice are:
>
> a. Intrahepatic cholestasis. This may be due to cirrhosis or hepatoma.
>
> b. Extrahepatic obstruction due to stones in the gallbladder or biliary tract; carcinoma of head of pancreas or enlarged lymph glands in the porta hepatis.

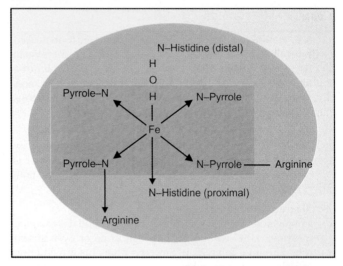

Fig. 22.10. Linkage of heme with globin. Pink circle represents the globin chain. Blue rectangle represents the protoporphyrin ring. The adult Hb (HbA) has a molecular weight of 67,000 D. It has 2 alpha chains and 2 beta chains. Hb F (fetal Hb) is made up of 2 alpha and 2 gamma chains. Hb A2 has 2 alpha and 2 delta chains. Normal adult blood contains 97% HbA, about 2% HbA2 and about 1% HbF. Alpha chain has 141 amino acids. The beta, gamma and delta chains have 146 amino acids. There are 4 heme residues per Hb molecule. The heme groups account for 4% of the mass of Hb. Linkage of heme with Iron: The iron in heme is hexavalent. It lies in the center of the tetrapyrole porphyrin ring, bonded to the four nitrogen atoms of the pyrole

Contd...

Contd...

rings. Of the remaining two co-ordinate bonds, the fifth co-ordination site is occupied by the proximal histidine residue of the alpha (87th amino acid) or beta (92th amino acid) chains. The sixth valancy binds oxygen in oxyhemoglobin .This oxygen atom directly binds to iron and forms a hydrogen bond with the imidazole nitrogen of the distal histidine residues of alpha (58) or beta (63) chain. The heme is located in a hydrophobic pocket in the cleft between the E and F helices of the globin chains. The propionyl substituent group of the porphyrin ring forms electrostatic bonds with two separate arginine residues of the globin

Table 22.6. Oxygenation and oxidation

When hemoglobin carries oxygen, the Hb is **oxygenated**. The iron atom in Hb is still in the ferrous state. The oxygen atom directly binds to iron atom, and forms a hydrogen bond with an imidazole nitrogen of the distal histidine. In deoxy-Hb, a water molecule is present between the iron and distal histidine (Fig. 22.10).

Oxidised hemoglobin is called Met-Hb; then iron is in ferric state and the oxygen carrying capacity is lost.

Table 22.7. The Bohr effect and 2,3-BPG

The influence of pH and pCO_2 to facilitate oxygenation of Hb in the lungs and deoxygenation at the tissues is known as the Bohr effect. Binding of CO_2 forces the release of O_2. When the pCO_2 is high, CO_2 diffuses into the red blood cells, and forms carbonic acid (H_2CO_3). When carbonic acid ionizes, the intracellular pH falls.

The 2,3-BPG concentration is higher in young children compared to the elderly. The 2,3-BPG, preferentially binds to deoxy-Hb. During oxygenation, BPG is released (Fig.15.11). The high oxygen affinity of fetal blood (HbF) is due to the inability of gamma chains to bind 2,3-BPG.

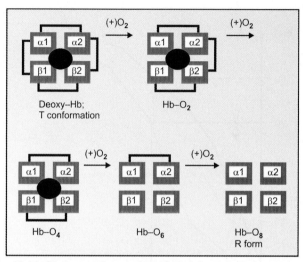

Fig. 22.11. Diagrammatic representation of the subunit interaction in hemoglobin. Pink rectangles represent Hb monomers. Black connection lines represent salt bridges. As oxygen is added, salt bridges are successively broken and finally 2,3-BPG is expelled. Simultaneously the T (taught) confirmation of deoxy-Hb is changed into R (relaxed) confirmation of oxy-Hb. Blue circle represents 2,3-bisphospho glycerate (BPG). When hemoglobin is in the deoxygenated state it exists in a tight (T) conformation with the salt bridges intact. When oxygen binds to one of the subunits, the salt bridges are broken and this would favor further oxygen binding to the other subunits (positive co-operative effect). As hemoglobin is fully oxygenated, 2,3-BPG is expelled from its position between the beta chains. Then the oxyhemoglobin assumes the relaxed (R) conformation. The reverse occurs when deoxygenation occurs with hemoglobin reverting back to the tight conformation. This also occurs with a co-operative interaction—release of oxygen from one subunit favoring dissociation of oxygen from other subunits

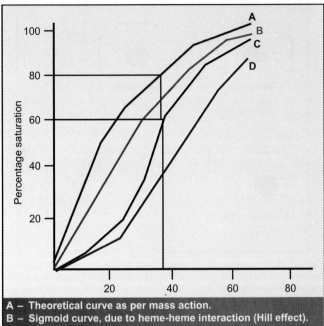

A – Theoretical curve as per mass action.
B – Sigmoid curve, due to heme-heme interaction (Hill effect).
C – Further shift to right due to carbon dioxide (Bohr effect) and BPG. This curve represents the pattern under normal conditions.
D – Further shift to right when temperature is increased to 42°C.

Fig. 22.12. Oxygen dissociation curve (OCD) of hemoglobin.

Contd...

Contd...

The Oxygen dissociation curve (ODC) of hemoglobin has a sigmoid shape because of the co-operative effect. It is seen that hemoglobin is more than 95% saturated with oxygen in the lung alveoli, where the PO_2 is high. In the tissue capillaries where the PO_2 is only 40 mms of Hg, hemoglobin is only 60% saturated. This would ensure adequate oxygen delivery to the tissues . 40% of the oxygen carried by 15 gm of Hb present in 100 ml of blood which amounts to 20 ml oxygen. The ODC shifts to the right liberating more oxygen when the pH is low (Bohr effect), pCO_2 is high and 2,3-BPG level is high. Oxygen remains bound and ODC shifts to the left at higher pH, low PCO_2 and when 2,3-BPG level falls

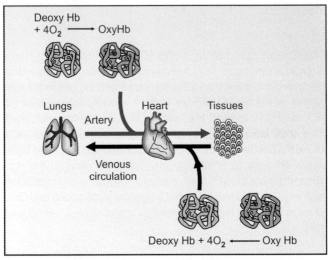

Fig. 22.13. In tissues oxy-Hb releases oxygen

Table 22.8. Heme-heme interaction

The sigmoid shape of the oxygen dissociation curve (ODC) is due to the allosteric effect, or co-operativity.

The binding of oxygen to one heme residue increases the affinity of remaining heme residues for oxygen (Fig. 15.12B). This is called **positive** co-operativity.

$$\underset{(+)O_2}{Hb \longrightarrow} \underset{(+)O_2}{HbO_2 \longrightarrow} \underset{(+)O_2}{HbO_4 \longrightarrow} \underset{(+)O_2}{HbO_6 \longrightarrow} HbO_8$$

| Affinity 1 time | Affinity 2 times | Affinity 4 times | Affinity 18 times |

Table 22.9. Transport of carbon dioxide

1. **Dissolved Form**

   About 10% of CO_2 is transported as dissolved form.

   $$CO_2 + H_2O \rightarrow H_2CO_3 \rightarrow HCO_3^- + H^+$$

   The hydrogen ions thus generated, are buffered by the buffer systems of plasma.

2. **Isohydric Transport of Carbon Dioxide**

   i. Isohydric transport constitutes about 75% of CO_2. It means that there is minimum change in pH during the transport. The H^+ ions are buffered by the deoxy-Hb and this is called the **Haldane effect.**

   ii. **In tissues:** Inside tissues, pCO_2 is high and carbonic acid is formed. It ionizes to H^+ and HCO_3^- inside the RBCs. The H^+ ions are buffered by deoxy-Hb and the HCO_3^- diffuses out into the plasma. Thus the CO_2 is transported from tissues to lungs, as plasma bicarbonate, without significant lowering of pH. The H^+ are bound by N-terminal NH_2 groups and also by the imidazole groups of **histidine** residues.

   iii. **Oxy-Hb is More Negatively Charged than Deoxy-Hb:** The iso-electric point of oxy-hemoglobin is 6.6, while that of deoxy-Hb is 6.8. Thus oxy Hb is more negatively charged than deoxy Hb. The reaction in tissues may be written as :

   $$OxyHb^= + H^+ \rightarrow HHb^- + O_2$$

   Therefore some cation is required to remove the extranegative charge of Oxy-Hb. So H^+ are trapped. 1 millimol of deoxy Hb can take up 0.6 mEq of H^+.

Contd...

Contd...

iv. **In the lungs:** In lung capillaries, where the pO_2 is high, oxygenation of hemoglobin occurs. When 4 molecules of O_2 are bound and one molecule of hemoglobin is fully oxygenated, hydrogen ions are released.

$$H{-}Hb + 4O_2 \longrightarrow Hb(O_2)_4 + H^+$$

v. The protons released in the RBC combine with HCO_3^- forming H_2CO_3 which would dissociate to CO_2, that is expelled through pulmonary capillaries.

3. **Carriage as Carbamino-Hemoglobin**

The rest 15% of CO_2 is carried as carbamino-hemoglobin, without much change in pH. A fraction of CO_2 that enters into the red cell is bound to Hb as a carbamino complex.

$$R{-}NH_2 + CO_2 \longrightarrow R{-}NH{-}COOH$$

The N-terminal amino group (valine) of each globin chain forms carbamino complex with carbon dioxide.

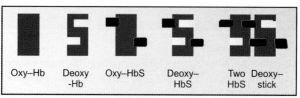

Fig. 22.14. Sticky patches on HbS molecule

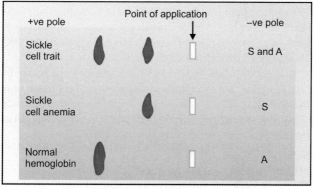

Fig. 22.15. Electrophoresis at pH 8.6

Table 22.10. Hemoglobinopathy and thalassemia

Abnormalities in the primary sequence of globin chains lead to **hemoglobinopathies,** e.g. HbS. Abnormalities in the rate of synthesis would result in **thalassemias.** In other words, normal hemoglobins in abnormal concentrations result in thalassemias, e.g., beta thalassemia

Table 22.11. Hemoglobin S (HbS)

Sickle Cell Hemoglobin

Of the hemoglobin variants, HbS constitutes the **most common** variety worldwide. In 1949 Linus Pauling (Nobel prize, 1954) established that HbS has abnormal electrophoretic mobility.

Sickle Cell Disease

The glutamic acid in the **6th position** of beta chain of HbA is changed to valine in HbS. This single amino acid substitution leads to polymerisation of hemoglobin molecules inside RBCs. This causes a distortion of cell into sickle shape.

The substitution of hydrophilic glutamic acid by hydrophobic valine causes a localised stickiness on the surface of the molecule (Fig. 22.14). The sickling occurs under deoxygenated state. The sickled cells form small plugs in capillaries. Death usually occurs in the second decade of life.

Sickle Cell Trait

In heterozygous (AS) condition, 50% of Hb in the RBC is normal. Therefore the sickle cell trait as such does not produce clinical symptoms. Such persons can have a normal life span.

At higher altitudes, **hypoxia** may cause manifestation of the disease. Chronic **lung disorders** may also produce hypoxia-induced sickling in HbS trait.

Table 22.12. Thalassemias

The name is derived from the Greek word, "thalassa", which means "sea". Greeks identified this disease present around Mediterranean sea. Thalassemia may be defined as the normal hemoglobins in abnormal proportions .

Reduction in alpha chain synthesis is called alpha thalassemia, while deficient beta chain synthesis is the beta thalassemia.

1. **Beta thalassemia**
   Beta thalassemia is more common than alpha variety. Beta type is characterised by a decrease or absence of synthesis of beta chains. As a compensation, gamma or delta chain synthesis is increased.

2. **Alpha thalassemias**
   Alpha thalassemia is rarer because alpha chain deficiency is incompatible with life.

3. **Thalassemia syndromes**
   All cases of thalassemias are characterised by **deficit of HbA** synthesis. Hypochromic microcytic anemia is seen. In homozygous state, clinical manifestations are severe, and hence called **Thalassemia major**, e.g. **Cooley's anemia**. In heterozygous conditions, the clinical signs and symptoms are minimal; they are called Thalassemia **minor**.

Table 22.13. Causes of anemia

Normal value for Hb in normal male is 14 to 16 g/dl and in female 13 to 15 g/dl.

1. **Anemias due to impaired production of RBCs**
   a. **Defect in heme synthesis:** This may be due to deficiency of nutritional factors such as **iron**, copper, pyridoxal phosphate, folic acid, vitamin B_{12} or vitamin C. Lead will inhibit heme synthesis.
   b. **Defect in stem cells:** Aplastic anemia due to drugs (e.g. Chloramphenicol), infections and malignant infiltrations may lead to anemia.

2. **Hemolysis due to intracorpuscular defect**
   a. **Hemoglobinopathies** such as HbS, HbC
   b. **Thalassemias**—major and minor
   c. **Enzyme deficiencies:** Deficiency of glucose-6-phosphate dehydrogenase.

3. **Hemolysis due to extracorpuscular causes**
   a. **Infections:** Malarial parasites.
   b. **Autoimmune hemolysis**
   c. **Hemolysis due to drug sensitization:** Many drugs including alpha-methyl dopa, quinine, etc.

4. **Hemorrhage**
   Hematuria, hematemesis, hemoptysis, peptic ulcer metrorrhagia and hemorrhoides are the usual causes for hemorrhage. Hemophilia (absence of AHG) and thrombocytopenia are other causes.

Section 7

Nutrition

23. Lipid Soluble Vitamins

Table 23.1. Comparison of two types of vitamins

	Fat soluble vitamins	Water soluble vitamins
Solubility in fat	Soluble	Not soluble
Water solubility	Not soluble	Soluble
Absorption	Along with lipids requires bile salts	*Absorption simple
Carrier proteins	Present	*No carrier proteins
Storage	Stored in liver	*No storage
Deficiency	Manifests only when stores are depleted	*Manifests rapidly as there is no storage
Toxicity	Hypervitaminosis may result	Unlikely, since excess is excreted
Major vitamins	A,D,E and K	B and C

*Vitamin B_{12} is an exception

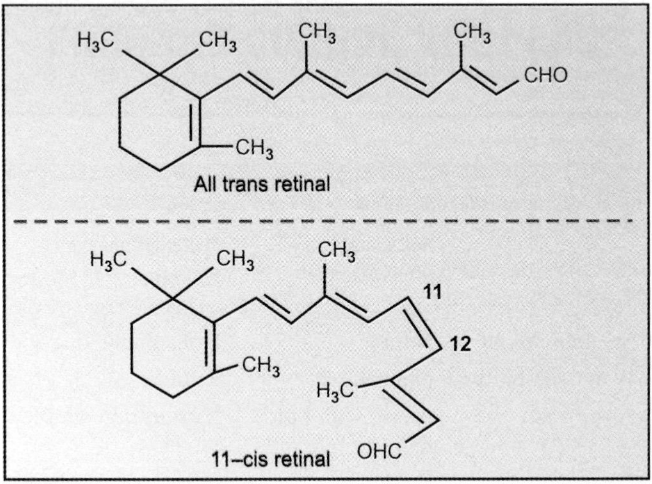

Fig. 23.1. Structure of vitamin A

Beta carotene present in food is cleaved by a dioxygenase using molecular oxygen to form retinal.Retinal may be reduced to retinol by a reductase.Retinol is esterified with fatty acids and present as retinyl ester. After hydrolysis the retinol is absorbed by the mucosal cell along with products of digestion of fat with the help of bile salts. In the mucosal cell it is re-esterified to form retinyl ester, incorporated into chylomicrons and transported to liver

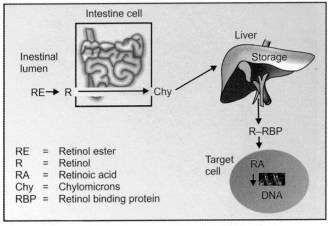

Fig. 23.2. Vitamin A metabolism

In the liver vitamin is stored as retinol palmitate. Transport to peripheral tissues is by retinol binding protein. The retinol-RBP complex binds to specific recptors on target cells, the vitamin binds to a cellular retinoic acid binding protein and is taken up. Effect is produced by binding to response elements in the cell.

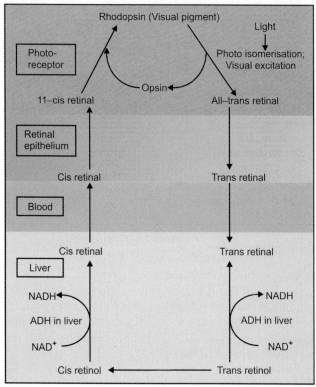

Fig. 23.3. Wald's visual cycle. Blue color represents reactions in photoreceptor matrix. Green background represents reactions in retinal pigment epithelium. Red depicts blood. Yellow shows reactions in liver

Table 23.2. Wald's visual cycle

The action of retinal in vision is explained by the Wald's visual cycle. Rhodopsin, a memebrane protein found on the photoreceptors of retina which has opsin and 11 cis retinal as the prosthetic group. When light falls on the retina, the 11 cis retinal isomerizes to 11 trans retinal causing visual excitation. The visual pigments act through a G protein coupled receptor—transducin. The photoexcitation generates cyclic GMP. It acts as a gate and the resultant closure of ion channels decrease sodium influx causing light induced hyperpolarisation. The nerve impulse thus generated in the retina is transmitted to visual centres in the brain. Wald's visual cycle regenerates 11 cis retinal.

Photon induced isomerisation of 11 cis retinal leads to dissociation of all trans retinal from rhodopsin. It can be directly isomerised to 11 cis retinal in the retina itself by an isomerase in the dark. This is a slow process of regeneration of rhodopsin. Most of the trans retinal is transported to the liver, reduced to all trans retinol by an NADH dependent alcohol dehydrogenase. The all transretinol is isomerised to 11 cis retinal and then oxidsed to 11 cis retinal which is then transported to retina completing the visual cycle.

Table 23.3. Biochemical functions of vitamin A

Retinal is the active form required for normal vision.
Retinoic acid is implicated in growth and **differentiation** of tissues. Retinol is necessary for normal **reproduction**. Vitamin A has good **anti-oxidant property.**

Rods are for vision in dim light
In the retina, there are two types of photosensitive cells, the rods and the cones. Rods are responsible for perception in dim light. It is made up of 11-cis-retinal + opsin. Deficiency of cis-retinal will lead to night blindness.

Cones are for color vision
Cones are responsible for vision in bright light as well as color vision. They contain the photosensitive protein, **conopsin** (photopsin). In cone proteins also, 11-cis-retinal is the chromophore. Reduction in number of cones or the cone proteins, will lead to **color blindness.**

Dark adaptation mechanism
Bright light depletes stores of rhodopsin in rods. Therefore when a person shifts suddenly from bright light to a dimly lit area, there is difficulty in seeing. After a few minutes, rhodopsin is resynthesised and vision is improved. This period is called **dark adaptation time.** It is increased in vitamin A deficiency.

Table 23.4. Deficiency manifestations of vitamin A

Night blindness or nyctalopia: Visual acuity is diminished in dim light. The patient cannot read or drive a car in poor light. The dark adaptation time is increased.

Xerophthalmia: The conjunctiva becomes dry, thick and wrinkled. The conjunctiva gets keratinised and loses its normal transparency. Cornea is also keratinised. Infections may supersede.

Bitot's spots: These are seen as grayish-white triangular plaques firmly adherent to the conjunctiva. This is due to increased thickness of conjunctiva in certain areas. All the ocular changes mentioned so far are completely reversible when vitamin is supplemented.

Keratomalacia: When the xerophthalmia persists for a long time, it progresses to keratomalacia (softening of the cornea). Later, corneal opacities develop. Bacterial infection leads to corneal ulceration, and total blindness.

Skin and mucous membrane lesions
Hyperkeratosis of the epithelium occurs. Epithelium is **atrophied**. The alterations in skin may cause increased occurrence of generalised **infections**. Therefore in old literature, vitamin A is referred to as anti-inflammatory vitamin.

Table 23.5. Nutritional aspects of vitamin A

Causes for vitamin A deficiency
Decreased intake. Obstructive jaundice causing defective absorption. Chronic nephrosis, where RBP is excreted through urine.

Dietary sources of vitamin A
Animal sources include milk, butter, cream, cheese, egg yolk and liver. Fish liver oils (cod liver oil and shark liver oil) are very rich sources of the vitamin. Vegetable sources contain the yellow pigment beta carotene. **Carrot** contains significant quantity of beta carotene. **Papaya, mango,** pumpkins, green leafy vegetables (spinach, amaranth) are other good sources for vitamin A activity.

Daily requirements of vitamin A
The recommended daily allowance (RDA) for
Children = 400-600 microgram.
Men = 750-1000 microgram/day
Women = 750 microgram/day
Pregnancy = 1000 microgram/day

Hypervitaminosis A or vitamin toxicity
It has been reported in children where parents have been overzealous in supplementing the vitamins. Symptoms of toxicity include anorexia, irritability, headache, peeling of skin, drowsiness and vomiting. Enlargement of liver is also seen.

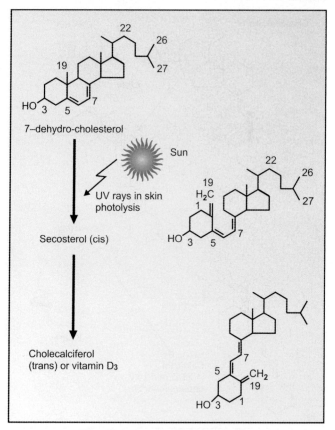

Fig. 23.4. Synthesis of vitamin D$_3$

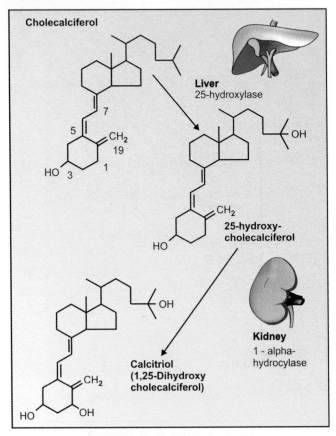

Fig. 23.5. Generation of calcitriol

Table 23.6. Calcitriol and calcitonin are different

Calcitriol is the physiologically active form of vitamin D. It increases the blood calcium level. Calcitriol acts like a steroid hormone. Being lipophilic, it passes into the cell, combines with a cytoplasmic receptor. The complex translocates into the nucleus, combines with a binding site on DNA causing transcription of mRNA for calbindin, a calcium binding protein. Calcium is absorbed from the intestine.

Calcitonin is the peptide hormone released from thyroid gland. It decreases the blood calcium.

Table 23.7. Actions of vitamin D

Effect on absorption of calcium
Calcitriol promotes the absorption of calcium and phosphorus from the intestine.

Effect of vitamin D in bone
Mineralisation of the bone is increased by increasing the activity of osteoblasts. Calcitriol stimulates osteoblasts which secrete alkaline phosphatase. Due to this enzyme, the local concentration of phosphate is increased. The ionic product of calcium and phosphorus increases, leading to mineralisation.

Table 23.8. Nutrional aspects of vitamin D

Deficiency of vitamin D
The deficiency diseases are **rickets** in children and **osteomalacia** in adults. Hence vitamin D is known as antirachitic vitamin. The classical features of rickets are **bone deformities**. Weight bearing bones are bent. The abnormalities in **biochemical parameters** are a slightly lower serum calcium, and a low serum phosphate. Serum **alkaline phosphatase** is markedly increased. It may be noted that vitamin D deficiency never produces severe hypocalcemia. Tetany will not be manifested.

Causes for vitamin D deficiency
Nutritional deficiency of vitamin D can occur in people who are not exposed to sunlight properly, e.g. in northern latitudes, in winter months.

Malabsorption of vitamin (obstructive jaundice and steatorrhea). Abnormality of vitamin D activation. Liver and renal diseases may retard the hydroxylation reactions.

Requirements of vitamin D
Children = 10 microgram (400 IU)/day
Adults = 5 to 10 microgram (200 IU)/day
Pregnancy, lactation = 10 microgram/day

Sources of vitamin D
Exposure to **sunlight**. Fish liver oil, fish and egg yolk are good sources of the vitamin.

Table 23.9. Nutrional aspects of vitamin E

Chemical nature
They have a chromane ring (tocol) system, with an isoprenoid side chain. There are eight naturally occurring tocopherols. Of these, **alpha tocopherol** has greatest biological activity.

Biochemical role of vitamin E
Vitamin E is the most powerful natural anti-oxidant
Vitamin E protects RBC from hemolysis. Vitamin E prevents early aging. It reduces the risk of myocardial infarction by reducing oxidation of LDL

Inter-relationship with selenium
Selenium has been found to decrease the requirement of vitamin E and vice versa. They act synergistically to minimise lipid peroxidation.

Deficiency manifestations of vitamin E
Human deficiency has not been reported. But in volunteers, vitamin E deficiency has been shown to produce muscular weakness.

Sources of vitamin E
Vegetable oils are rich sources of vitamin E, e.g. wheat germ oil, sunflower oil, safflower oil, cotton seed oil, etc.

Requirement
About 10 mg per day. The requirement increases with higher intake of PUFA.

Table 23.10. Nutrional aspects of vitamin K

1. Chemistry of vitamin K
The letter "K" is the abbreviation of the German word "koagulation vitamin". They are **naphtho-quinone** derivatives, with a long isoprenoid side chain. **Menadione** is water-soluble synthetic vitamin.

2. Biochemical role of vitamin K
Vitamin K is necessary for coagulation. Factors such as Factor II (**prothrombin**) and Factor IX (Christmas factor) and Factors VII and IX. The **gamma carboxy glutamic acid** (GCG) synthesis requires vitamin K as a co-factor (Fig. 23.6).

Causes for deficiency of vitamin K
Deficiency can occur in conditions of **malabsorption** of lipids. This can result from obstructive jaundice. Prolonged **antibiotic** therapy will destroy the bacterial flora and can lead to vitamin K deficiency.

Clinical manifestations of deficiency
It is manifested as **bleeding**. Prolongation of prothrombin time and delayed clotting time are seen. **Warfarin** and **dicoumarol** will competitively inhibit the gamma carboxylation system due to structural similarity with vitamin K. Hence, they are widely used as anticoagulants for therapeutic purposes.

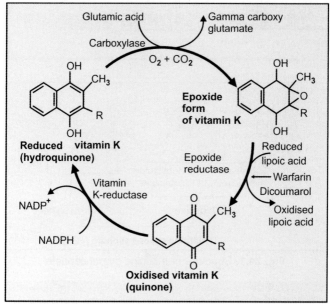

Fig. 23.6. Vitamin K cycle. The coagulation factors undergo post translational modifications (gamma carboxylation of glutamic acid residues) by a vitamin K dependent carboxylase which uses molecular oxygen. The reduced hydroquinone form of vitaminK is oxidized to an epoxide. The epoxide in turn is reduced to the quinine form using reduced lipoamide. The hydroquinone form is regenerated by vitamin K reductase which is an NADPH dependent enzyme

24. Water Soluble Vitamins

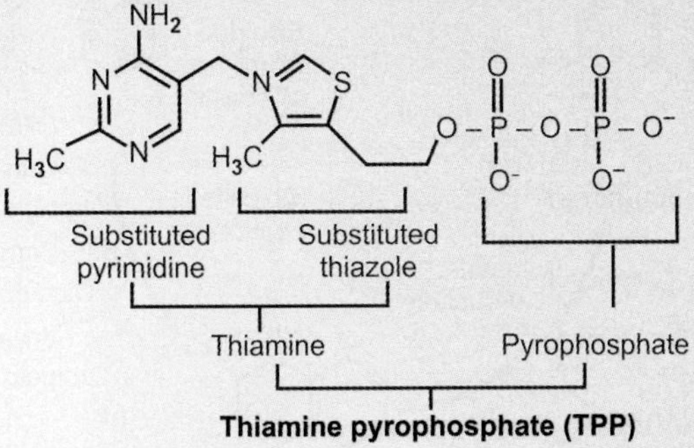

Fig. 24.1. Structure of thiamine pyrophosphate

Table 24.1. Nutrition of thiamine (vitamin B$_1$)

Sources

Cereals (whole wheat flour and unpolished handpound rice) are rich sources of thiamine. Yeast is also a very good source.

Physiological role of thiamine

i **Pyruvate dehydrogenase:** The co-enzyme form is **thiamine pyrophosphate** (TPP). It is used in oxidative decarboxylation of alpha keto acids, e.g. pyruvate decarboxylase, a component of the pyruvate

Contd...

Contd...

> dehydrogenase complex. It catalyses the breakdown of pyruvate, to acetyl CoA, and carbon dioxide.
> ii. **Alpha keto glutarate dehydrogenase:** in TCA cycle)
> iii. **Transketolase:** It is an enzyme in the hexose mono-phosphate shunt pathway of glucose.
>
> *Deficiency manifestations of thiamine*
> **Beriberi:** Deficiency of thiamine leads to beriberi.
>
> **Wet beriberi:** Here cardiovascular manifestations are prominent. Edema of legs, face, and serous cavities are the main features.
>
> **Dry beriberi:** Peripheral neuritis with sensory disturbance leads to complete paralysis.
>
> **Wernicke-Korsakoff syndrome:** Encephalopathy plus psychosis.
>
> **Polyneuritis:** It is common in chronic alcoholics, pregnancy and old age.
>
> *Recommended daily allowance of thiamine*
> It depends on calorie intake (0.5 mg/1000 calories). Requirement is 1-1.5 mg/day. Thiamine is useful in the treatment of beriberi, alcoholic polyneuritis, neuritis of pregnancy and neuritis of old age. The main role of thiamine (TPP) is in **carbohydrate** metabolism. So, the requirement of thiamine is increased along with higher intake of carbohydrates.

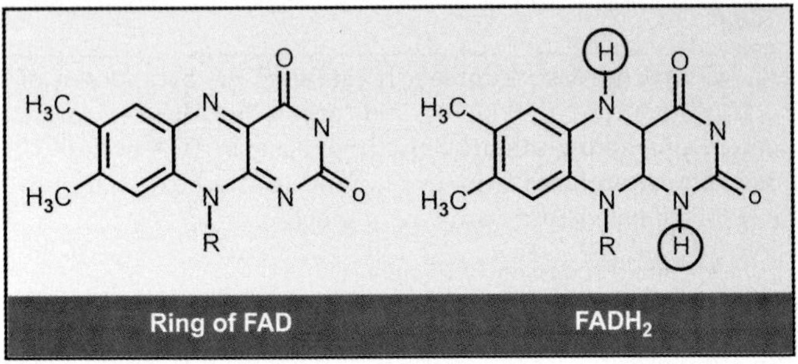

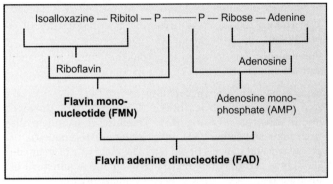

Fig. 24.2. Acceptance of hydrogen by FAD

Fig. 24.3. Co-enzymes FMN and FAD

Table 24.2. Nutrition of Riboflavin (Vitamin B2)

Riboflavin deficiency

Manifestations: Symptoms are confined to skin and mucous membranes. **Glossitis** (Magenta colored tongue), **cheilosis**, **angular stomatitis** and circumcorneal vascularisation are main manifestations.

Dietary sources of riboflavin

Rich sources are liver, dried yeast, egg and whole milk. Good sources are fish, cereals, legumes.

Daily requirement

Riboflavin requirement is related to calorie intake. Adults on sedentary work require about 1.5 mg per day. During pregnancy, lactation and old age, additional 0.2 to 0.4 mg/day are required.

Table 24.3. FAD-dependent enzymes

1. Succinate to fumarate by succinate dehydrogenase.
2. Acyl CoA to alpha-beta unsaturated acyl CoA by acyl CoA dehydrogenase
3. Xanthine to uric acid by xanthine oxidase.
4. Pyruvate to acetyl CoA by pyruvate dehydrogenase.
5. Alpha ketoglutarate to succinyl CoA by alpha ketoglutarate dehydrogenase.

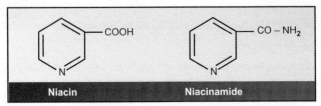

Fig. 24.4. Structure of niacin

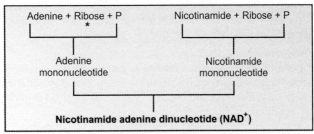

Fig. 24.5. Structure of NAD$^+$ (In NADP$^+$ phosphoric acid is attached to the ribose group marked with asterisk)

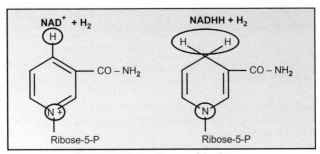

Fig. 24.6. Acceptance of hydrogen

Table 24.4. NAD⁺ dependent enzymes

1. Lactate dehydrogenase (lactate → pyruvate)
2. Glyceraldehyde-3-phosphate dehydrogenase
   (glyceraldehyde-3-phosphate → 1, 3-bisphosphoglycerate)
3. Pyruvate dehydrogenase
   (pyruvate → acetyl CoA)
4. Alpha ketoglutarate dehydrogenase
   (alpha ketoglutarate → succinyl CoA
5. Beta hydroxy acyl CoA dehydrogenase
   (beta hydroxy acyl CoA → beta keto acyl CoA
6. Glutamate dehydrogenase
   (Glutamate → alpha ketoglutarate)

Table 24.5. NADPH generating reactions

1. Glucose-6-phosphate dehydrogenase in the hexose
   monophosphate shunt pathway
   (Glucose-6-phosphate → 6-phospho-gluconolactone)
2. 6-phosphogluconate dehydrogenase in the shunt pathway
   (6-phosphogluconate → 3-keto-6-phosphogluconate)

Table 24.6. NADPH utilising reactions

1. Keto acyl ACP dehydrogenase (Beta keto acyl ACP → beta hydroxy acyl ACP)
2. Alpha, beta unsaturated acyl ACP → acyl ACP
3. HMG CoA reductase (HMG CoA → mevalonate
4. Met-hemoglobin → hemoglobin
5. Folate reductase (Folate → dihydrofolate → tetrahydro-folate)
6. Phenylalanine hydroxylase (Phenylalanine → tyrosine)

Table 24.7. Nutritional aspects of riboflavin

Niacin deficiency
Pellagra: Deficiency of niacin leads to the clinical condition called pellagra. The symptoms are:

  i. **Dermatitis:** Red erythema occurs, especially in the feet, ankles and face. Increased pigmentation around the neck is known as **Casal's necklace**.
 ii. **Diarrhea:** Diarrhea, nausea and vomiting.
iii. **Dementia:** It is frequently seen in chronic cases. Delerium is common in acute pellagra. Irritability, inability to concentrate and poor memory are more common in mild cases.

Niacin is synthesised from tryptophan
About 60 mg of tryptophan is equivalent to 1 mg of niacin.

Contd...

Contd...

Causes for niacin deficiency
   i. **Dietary deficiency of tryptophan:** Pellagra is seen among people whose staple diet is maize (South and Central America) or **sorghum** (jowar or guinea corn) as in central and western India.
  ii. **Lack of synthesis of vitamin B_6:** Kynureninase, an important enzyme in the pathway of tryptophan, is pyridoxal phosphate dependent. So conversion of tryptophan to niacin is not possible in pyridoxal deficiency.
 iii. **Isoniazid (INH):** This anti-tuberculous drug inhibits pyridoxal phosphate formation. Hence there is block in conversion of tryptophan to NAD^+
  iv. **Carcinoid syndrome:** The tumour utilises major portion of available tryptophan for synthesis of serotonin; so tryptophan is unavailable.

Dietary sources of niacin
Yeast, rice polishing, liver, peanut, whole cereals, legumes, meat and fish. About 60 mg of tryptophan will yield 1 mg of niacin.

Recommended daily allowance (RDA)
Normal requirement is 20 mg/day. During lactation, additional 5 mg are required.

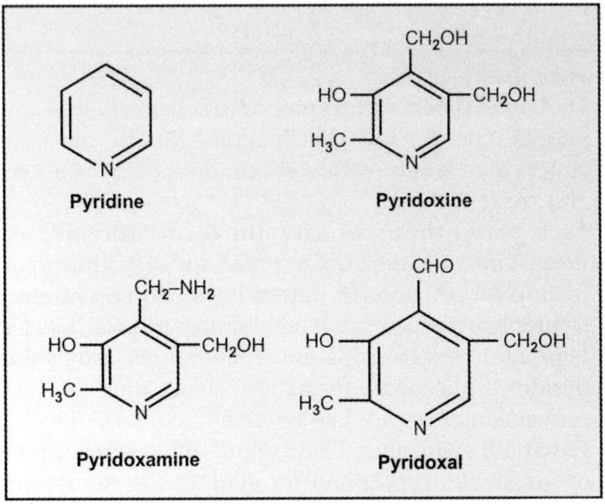

Fig. 24.7. Structure of B_6 related compounds

Table 24.8. Nutrition of pyridoxal (vitamin B_6)

Functions of pyridoxal phosphate

i. **Transamination:** These reactions are catalysed by amino transferases (transaminases) which employ PLP as the co-enzyme.

ii. **Decarboxylation:** All decarboxylation reactions of amino acids require PLP as co-enzyme. A few examples are given below:

    a. Glutamate → GABA (gamma aminobutyric acid)

    b. Histidine → histamine

Contd...

Contd...

iii. **Sulphur containing amino acids:** Methionine and
   cysteine metabolism.
   a. Homocysteine + Serine → Cystathionine. (Enzyme
      Cystathionine synthase)
   b. Cystathionine → Homoserine + Cysteine (Enzyme
      Cystathionase)
iv. **Heme synthesis:** ALA synthase is a PLP dependent
   enzyme. So, in B6 deficiency, **anemia** is common.
v. **Production of niacin:** Kynureninase is a PLP dependent
   enzyme. Hence in vitamin B_6 deficiency niacin
   production is less.
vi. **Glycogenolysis: Phosphorylase** enzyme (glycogen to
   glucose-1-phosphate) requires PLP.

Deficiency manifestations of pyridoxine
A. **Neurological:** Convulsions, demyelination of nerves and
   peripheral neuritis.
B. **Dermatological:** Since niacin is produced from
   tryptophan, B_6 deficiency in turn causes niacin deficiency
   which leads to **pellagra**.
C. **Hematological:** hypochromic microcytic **anemia**

Dietary sources
Yeast, rice polishing, wheat germs, cereals, pulses, oil seeds,
egg, milk, fish and green leafy vegetables.

Requirement of B_6
It is related to **protein intake** and not to calorie intake. Adults
need 1 to 2 mg/day. During pregnancy, the requirement is
increased to 2.5 mg/day.

Table 24.9. Nutritional aspects of biotin

Co-enzyme activity of biotin
Biotin acts as co-enzyme for **carboxylation reactions** (carbondioxide fixation reactions). In the first step, a molecule of CO_2 is captured by biotin forming carboxybiotin. Energy required for this reaction is provided by ATP. In the second step the activated carboxyl group is transferred to the substrate and biotinyl enzyme is regenerated.

Biotin requiring CO_2 fixation reactions
  i. **Acetyl CoA carboxylase:**
     Acetyl CoA $+CO_2+ATP\rightarrow$Malonyl CoA + ADP+Pi
 ii. **Propionyl CoA carboxylase:**
     Propionyl CoA $+CO_2+ATP\rightarrow$
     Methyl malonyl CoA +ADP+Pi
iii. **Pyruvate carboxylase:**
     Pyruvate + CO_2 +ATP$\rightarrow$Oxaloacetate +ADP +Pi

Biotin-independent carboxylation reactions
Carbamoyl phosphate synthetase, which is the stepping stone for urea and pyrimidine synthesis.

Biotin antagonists
Avidin, a protein present in **egg white** has great affinity to biotin. Hence intake of raw (unboiled) egg may cause biotin deficiency.

Requirement of biotin
About 200-300 mg per day.

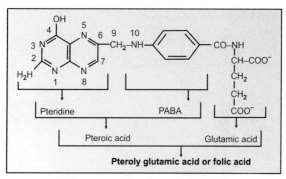

Fig . 24.8. Structure of folic acid

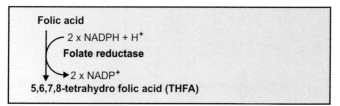

Fig . 24.9. Folate reductase

Table 24.10. Function of tetrahydro folic acid (THFA)

The THFA is the **carrier of one-carbon** groups. attached either to N_5 or N_{10} or in a cyclical structure between N_5 and N_{10}. The following groups are one carbon compounds: **i)** Formyl (–CHO); **ii)** Formimino (–CH = NH); **iii)** Hydroxy-methyl (–CH$_2$OH); **iv)** Methyl (–CH$_3$).

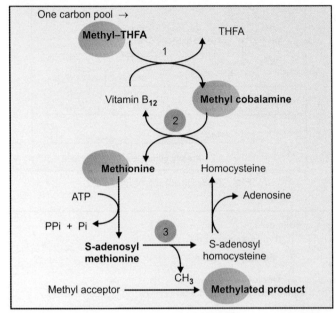

Fig. 24.10. Transmethylation reactions. (2) = Homocysteine methyl transferase (3) = methyl transferase. Once formed, N_5 methyl THFA can transfer its methyl group only to homocysteine, forming methionine (remethylation). This reaction requires B_{12} as coenzyme. The active methionine SAM can transfer the methyl group to acceptors to form methylated acceptors and S-adenosyl homocysteine. B_{12} deficiency traps THFA as N_5 methyl THFA ; this is called **folate trap**

Table 24.11. Nutritional aspects of Folic acid

Causes for folate deficiency
 i. **Pregnancy:** Folate deficiency is commonly seen in pregnancy, where requirement is increased.
 ii. **Drugs:** Anticonvulsant drugs (hydantoin, dilantin, phenytoin, phenobarbitone) will inhibit folate absorption.
iii. **Hemolytic anemias:** As requirement of folic acid becomes more, deficiency is manifested.
iv. **Dietary deficiency:** Absence of vegetables in food for prolonged periods.

Deficiency manifestations
Reduced DNA synsthesis: Thymidylate synthase enzyme is inhibited. So dTTP is not available for DNA synthesis. Thus cell division is arrested. Very rapidly dividing cells in bone marrow and intestinal mucosa are therefore most seriously affected.
Macrocytic anemia: It is the most characteristic feature of folate deficiency. **Immature looking nucleus** with mature eosinophilic cytoplasm in the bone marrow cells. **Reticulocytosis** is often seen. These abnormal RBCs are rapidly destroyed in spleen. This **hemolysis** leads to anaemia. Folic acid deficiency may cause increased homocystein levels in blood with increased risk of coronary artery diseases.

Sources of folic acid
Rich sources of folate are yeast, green leafy vegetables. Moderate sources are cereals, pulses, oil seeds and egg.

Contd...

Contd...

> *Recommended daily allowance (RDA)*
> The requirement of free folate is 200 microgram/day. In pregnancy the requirement is increased to 400 microgram/day and during lactation to 300 microgram/day.
>
> *Folate antagonists*
> **a. Sulphonamides**
>   They have structural similarity with PABA. Bacteria can synthesise folic acid from the components, pteridine, PABA and glutamate. When sulphonamides are given, micro-organisms cannot synthesise folic acid and hence their growth is inhibited. Thus sulphonamides are very good **antibacterial** agents, which do not affect the human cells.
> **b. Aminopterin and amethopterin**
>   Aminopterin (4-amino folic acid) and amethopterin (methotrexate) (4-amino, 10-methyl folic acid) are powerful inhibitors of folate reductase and THFA generation. Thus these drugs decrease the DNA formation and cell division. They are widely used as **anticancer** drugs, especially for leukemias and choriocarcinomas.

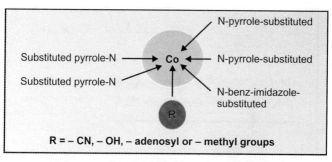

Fig. 24.11. Simplified structure of vitamin B$_{12}$

Table 24.12. Absorption of Vitamin B$_{12}$

Vitamin B$_{12}$ combines with the **intrinsic factor** (IF) of Castle. Hence the B$_{12}$ is otherwise known as extrinsic factor (EF), that is, the factor derived from external sources. Intrinsic factor is secreted by the gastric parietal cells. This IF-B$_{12}$ complex is attached with specific receptors on mucosal cells. The whole IF-B$_{12}$ complex is internalised.

Transport and storage

In the blood, **methyl B$_{12}$** form is predominant. **Transcobalamin**, is the specific carrier. It is stored in the liver cells, as **ado-B$_{12}$** form, in combination with **transcorrin**. Generally, B complex vitamins are not stored in the body, B$_{12}$ is an exception.

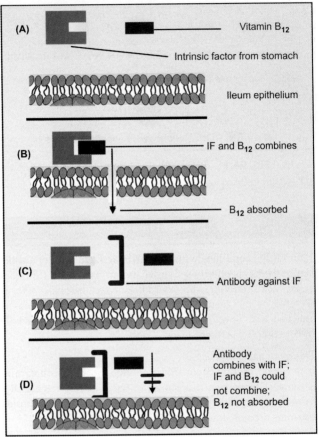

Fig. 24.12. Absorption of vitamin B$_{12}$.

Contd...

Contd... Legend of Figure 24.12. (A) Intrinsic factor secreted from stomach reaches intestine. (B) = Vitamin B$_{12}$ absorbed with the help of intrinsic factor. (C) = In pernicious anemia, antibody against IF is produced. (D) = In presence of antibody, absorption is not taking place.

The absorption of B$_{12}$ requires binding with intrinsic factor produced in the stomach. The absorption of B$_{12}$ mainly occurs at the ileum. In pernicious anemia, which is an autoimmune disorder, antibodies to the intrinsic factor prevent binding of B$_{12}$. So deficiency results.

Table 24.13. Functional Role of B$_{12}$

Methyl malonyl CoA isomerase
During the metabolism of odd chain fatty acids, the propionyl CoA is carboxylated to methyl malonyl CoA. It is then isomerised by methyl malonyl isomerase or mutase (containing Ado–B$_{12}$) to succinyl CoA, which enters into citric acid cycle. In B$_{12}$ deficiency, methyl malonyl CoA is excreted in urine (**methyl malonic aciduria**).

Homocysteine methyl transferase
See Step 2 in Figure 24.10. This enzyme needs vitamin B$_{12}$ (methyl cobalamin).

Methyl folate trap and folate deficiency
See note under Figure 24.10.

Table 24.14. Nutritional aspects of B$_{12}$

Causes of B$_{12}$ deficiency
1. **Nutritional.**
2. **Decrease in absorption**
3. **Addisonian pernicious anemia:** It is an autoimmune disease. Antibodies are generated against IF (Figs 24.12C and D).
4. **Pregnancy:** Increased requirement

Deficiency manifestations
  i. **Folate trap:** See note under Figure 24.10.
 ii. **Megaloblastic anemia**
iii. **Abnormal homocysteine level:** Step No. 2 (Fig. 24.10) is blocked, leading to homocysteinemia and **homocystinuria**. Homocysteine level in blood is related with myocardial infarction.
iv. **Subacute combined degeneration:** Damage to nervous system is seen in B$_{12}$ deficiency (but not in folate deficiency). There is **demyelination** affecting cerebral cortex as well as dorsal column and pyramidal tract of spinal cord. Since sensory and motor tracts are affected, it is named as combined degeneration.

Requirement of vitamin B$_{12}$
Normal daily requirement is 1-2 microgram/day.

Dietary sources
Liver is the richest source. Curd is a good source.

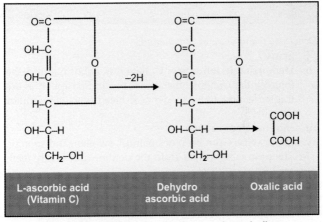

Fig. 24.13. Vitamin C: structure and catabolism

Table 24.15. Biochemical functions of vitamin C

i. **Hydroxylation of proline and lysine in collagen:** This process is necessary for the normal production of supporting tissues such as osteoid, collagen and intercellular cement substance of capillaries.

ii. **Iron metabolism:** Ascorbic acid enhances the iron absorption from the intestine.

iii. **Hemoglobin metabolism:** It is useful for re-conversion of met-hemoglobin to hemoglobin.

iv. **Anti-oxidant property**.

Table 24.16. Nutritional aspects of vitamin C

Deficiency manifestations of vitamin C
  i. **Scurvy**
 ii. **Hemorrhagic tendency:** Collagen is abnormal and the intercellular cement substance is brittle. So capillaries are fragile, leading to the tendency to bleed even under minor pressure.
iii. **Internal hemorrhage.**
 iv. **Oral cavity:** Gum becomes painful, swollen, and spongy. The pulp is separated from the dentine and finally teeth are lost.
  v. **Bones:** Failure of the osteoblasts to form the intercellular substance, osteoid. Without the normal ground substance, the deposition of bone is arrested. Scorbutic bone is weak and fractures easily.
 vi. **Anemia:** Microcytic, hypochromic anemia.

Dietary sources of vitamin C
Rich sources are amla (Indian gooseberry), lime, lemon and green leafy vegetables.

Requirement of vitamin C
Recommended daily allowance is 75 mg/day (equal to 50 ml orange juice). During pregnancy, lactation, and in aged people requirement may be 100 mg/day.

Therapeutic use of vitamin C
Vitamin C has been recommended for treatment of ulcer, trauma, and burns.

25. Mineral Metabolism

Table 25.1. Important minerals

Major elements	Trace elements
1. Calcium	1. Iron
2. Magnesium	2. Iodine
3. Phosphorus	3. Copper
4. Sodium	4. Manganese
5. Potassium	5. Zinc
6. Chloride	6. Molybdenum
7. Sulphur	7. Selenium
	8. Fluoride

Table 25.2. Selected list of Ca^{++} dependent enzymes

Activated by Ca^{++} and mediated by calmodulin	Activated directly by Ca^{++}
Adenyl cyclase	Pancreatic lipase
Phospholipase C	Enzymes of
Ca^{++} –Mg^{++} –ATPase	coagulation pathway
Glycogen synthase	
Myosin kinase	
Pyruvate dehydrogenase	
Ca^{++} dependent protein kinases	

Table 25.3. Functions of calcium

i. **Activation of enzymes:** See Table 25.2.

ii. **Muscles:** Calcium mediates **excitation and contraction** of muscle fibers. Upon getting the neural signal, calcium is released from sarcoplasmic reticulum. Calcium activates ATPase; increases reaction of actin and myosin and facilitates excitation-contraction coupling. Calcium decreases neuromuscular irritability. Calcium deficiency causes tetany.

iii. **Nerves:** Transmission of **nerve** impulses from pre-synaptic to post-synaptic regions.

iv. **Secretion of hormones:** Calcium mediates secretion of Insulin and parathyroid hormone.

v. **Second messenger:** Calcium and cyclic AMP are second messengers of different hormones. One example is glucagon.

vi. **Coagulation:** Calcium is known as factor IV in blood coagulation cascade.

vii. **Myocardium:** Ca^{++} **prolongs systole**. In hypercalcemia, cardiac arrest is seen in systole.

viii. **Bone and teeth:** The bulk quantity of calcium is used for bone and teeth formation. Bones also act as reservoir for calcium in the body.

Table 25.4. Factors regulating blood calcium

A. Vitamin D

  i. The active form of vitamin D is called dihydroxy-cholecalciferol or **calcitriol** (Fig. 23.5). Calcitriol and calcitonin are different (See below). The calcitriol induces a carrier protein in the intestinal mucosa, which increases the absorption of calcium.

 ii. In the bone, vitamin D increases the number and activity of **osteoblasts**, and increases the secretion of **alkaline phosphatase** by osteoblasts.

B. Parathyroid hormone (PTH)

  i. PTH acts through **cyclic AMP.**

 ii. In the **bone**, PTH causes demineralisation or decalcification. It induces pyrophosphatase in the **osteoclasts.** The number of osteoclasts are also increased. Osteoclasts release lactate into surrounding medium which solubilises calcium.

iii. In **kidney**, PTH causes decreased renal excretion of calcium and increased excretion of phosphates.

C. Calcitonin

  i. Calcitonin decreases serum calcium level. It **inhibits resorption of bone**. It decreases the activity of osteoclasts and increases that of osteoblasts.

 ii. Calcitonin and PTH are directly antagonistic. The PTH and calcitonin together promote the bone growth and remodeling.

Table 25.5. Comparison of action of three major factors affecting serum calcium

	Vitamin D	*PTH*	*Calcitonin*
Blood calcium	Increased	Drastically increased	Decreased
Main action	Absorption from gut	Demineral-isation	Opposes de-mineralisation
Calcium absorption from gut	Increased	Increased (indirect)	—
Bone resorption	Decreased	Increased	Decreased
Deficiency manifestation	Rickets	Tetany	—
Effect of excess	Hypercal-cemia$^+$	Hypercal-cemia^{++}	Hypocal-cemia

Table 25.6. Nutritional aspects of calcium

1. *Sources of calcium*: **Milk** is a good source for calcium. Egg, fish and vegetables are medium source for calcium.
2. *Daily Requirement of calcium*: *An adult needs 500 mg per day and a child about 1200 mg/day.*
3. *Hypercalcemia*: The term denotes that the blood calcium level is more than 11 mg/dl. The major cause is **hyper parathyroidism** Calcium may be precipitated in urine, leading to recurrent bilateral urinary **calculi.**
4. *Hypocalcemia*: When serum calcium level is less than 7.5 mg/dl, tetany, will result.

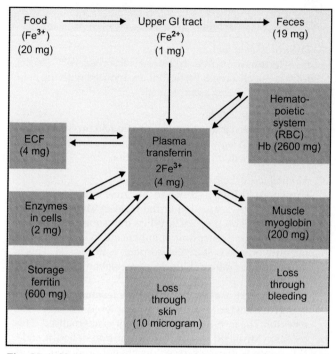

Fig. 25.1. Normal iron kinetics. Total body iron content is 3 to 5 gm; 75% of which is in blood. **Heme containing** proteins are hemoglobin, myoglobin, cytochromes, cytochrome oxidase, catalase. **Non-heme iron containing** proteins are transferrin, ferritin, hemosiderin. Blood contains **14.5 g of Hb per 100 ml**. About 75% of total iron is in hemoglobin, and 5% is in myoglobin and 15% in ferritin

Table 25.7. Factors influencing iron absorption

A. *Reduced form of iron*

   Fe^{++} (**ferrous**) form (reduced form) is absorbed. Fe^{+++} (ferric) form is not absorbed. Ferric ions are reduced with the help of gastric HCl, and ascorbic acid.

B. *Interfering substances*

   Iron absorption is decreased by **phytic acid** (in cereals) and **oxalic acid** (in leafy vegetables). Calcium, copper, lead and phosphates will inhibit iron absorption.

C. *Mucosal block theory*

   Iron metabolism is unique because homeostasis is maintained by regulation at the **level of absorption** and not by excretion (Fig. 25.2). When iron stores in the body are depleted, absorption is enhanced. When adequate quantity of iron is stored, absorption is decreased. This is referred to as "**mucosal block**" of regulation of absorption of iron.

   Iron enters the mucosal cell in the **ferrous state**. This is bound to **transferrin**. This is then complexed with a specific **receptor**. The iron-transferrin-receptor is internalised. This receptor mediated uptake is more in iron-deficient state. When iron is in excess, receptors are not produced; this is the basis of "mucosal block".

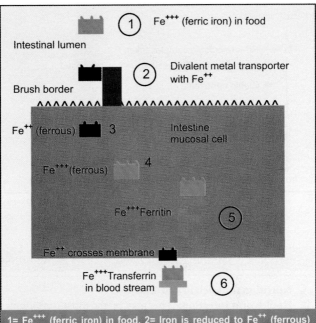

1= Fe^{+++} (ferric iron) in food. 2= Iron is reduced to Fe^{++} (ferrous) state, and attaches to divalent metal transporter on the mucosal surface. 3= Ferrous iron is internalised. 4= Iron is oxidised to ferric state. 5= Ferric iron binds with ferritin for temporary storage. 6= Ferric iron released, reduced to ferrous state by ceruloplasmin. In the bloodstream, ferric iron is bound with transferrin

Fig. 25.2. Absorption of iron from intestine

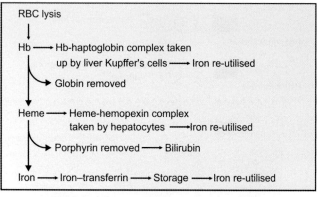

Fig. 25.3. Conservation of iron in the body

Table 25.8. Causes for iron deficiency anemia

i. **Nutritional deficiency of iron**
ii. **Hookworm infection:** This may be the most important cause, especially in rural areas, where sanitation is poor.
iii. **Repeated pregnancies:** About 1 g of iron is lost from the mother during one delivery.
iv. **Chronic blood loss:** Hemorrhoids (piles), peptic ulcer, uterine hemorrhage.

Iron deficiency is characterised by **microcytic hypochromic** anemia. Anemia results when hemoglobin level is less than 12 gm/dl.

Table 25.9. Nutritional aspects of iron

Requirement of iron (ICMR, 1990)

Daily allowance for iron for an adult Indian is **20 mg** of iron, out of which about 1-2 mg is absorbed. In Western countries, requirement is less (15 mg/day) because the diet does not contain inhibitory substances.

Pregnant women need 40 mg/day. Transfer of iron and calcium from mother to fetus occurs mainly in the last trimester of pregnancy. Therefore during this period mother's food should contain surplus quantities of iron and calcium.

In the first 3 months of life, iron intake is negligible because milk is a poor source of iron. During this time, child is dependent on the iron reserve received from mother during pregnancy. After 3 months of life, diet supplementation with cereals is essential for supplying the iron requirement.

Sources of iron

   i. **Leafy vegetables** are good sources.
  ii. Cereals contains moderate quantity of iron.
 iii. Liver and meat contain moderate quantities.
  iv. **Jaggery** is a good source for iron.
   v. Cooking in **iron vessels** will help to get iron.
 vii. Milk is a **very poor source** of iron.

Table 25.10. Lead poisoning

Causes of lead poisoning

Lead is the most common environmental poison It is dispersed into air, food, soil and water.

Paint is the major source for exposure, especially in children, as they bite painted toys. Paint is peeled off as small flakes from walls of living rooms.

Increased content of lead is seen in air, water and vegetables in cities and near highways. This is due to the tetraethyl lead derived from the **exhaust of vehicles**.

Lead pipes are important sources for contamination. **Battery repair**, radiator repair, soldering, painting and printing are occupations prone to get lead poisoning.

Newspapers and xerox copies contain lead, which is adsorbed to fingertips, and later contaminate foodstuff taken by hands.

One pack of **cigarette** contains 15 microgram of lead and chronic smokers have higher blood levels of lead.

Manifestations of lead poisoning

Lead is a **cumulative poison** and is accumulated in tissues over the years. More than 10 mg/dl in children and more than 25 mg/dl in adults leads to toxic manifestations. Lead is taken up from environment by enamel and **dentine**. Lead inhibits **heme synthesis**. Lead particularly inhibits delta amino levulinic acid (ALA) synthase and ALA-dehydratase

Miscarriage, still birth, and premature birth are reported in lead poisoning of mothers. Permanent **neurological** sequelae, cerebral palsy and optic atrophy may be seen. In children, **mental retardation**, learning disabilities, behavioral problems, hyperexcitability and seizures are seen. **Anemia**, abdominal colic and loss of appetite are very common.

Table 25.11. Summary of mineral metabolism

	Requirement for adult male/day	Blood level
Calcium	500 mg	9-11 mg/dl
Phosphorus	500 mg	3-4 mg/dl
Sodium	5-10 g	136-145 mEq/L
Potassium	3-4 g	3.5-5 mEq/L
Chloride	—	96-106 mEq/L
Iron	20 mg	120 µg/dl (plasma)
Copper	1.5 -3 mg	100 µg/dl
Iodide	150-200 µg	5-10 µg/dl
Zinc	15 mg	100 µg/dl

26. Energy Metabolism

Table 26.1. Energy yield from nutrients

Nutrients	Calorific values in kilocalories/g
Carbohydrates	4
Fats	9
Proteins	4
Alcohol	7

Table 26.2. Basal metabolic rate (BMR)

Definition: The basal metabolic rate is the energy required by an awake individual during physical, emotional and digestive rest.

Factors: Affecting BMR:

Age: In old age BMR is lowered.

Sex: Males have a higher BMR than females.

Temperature: BMR increases in cold climate.

Exercise: increases BMR

Fever: 12% increase in BMR is noticed per degree centigrade rise in temperature.

Thyroid hormones: Since thyroid hormones have a general stimulant effect on rate of metabolism and heat production, BMR is raised in hyperthyroidism and lowered in hypothyroidism.

Table 26.3. Specific dynamic action (SDA)

This refers to the increased heat production following the intake of food (**thermogenic effect** of food) (regulatory thermogenesis). This is due to the expenditure of energy for digestion and absorption of food.

This energy is trapped from previously available energy. SDA can be considered as the activation energy needed for a chemical reaction, to be supplied initially. Suppose a person takes 250 g of carbohydrates; this should produce $250 \times 4 = 1000$ kcal. But before this energy is trapped, about 10% energy (= 100 kcal) is drawn from the reserves of the body. Thus the net generation of energy is only 1000 minus 100 = 900 kcal. If the person wants to get 1000 kcal, he should take food worth 1100 kcal. Thus additional calories, equivalent to SDA has to be added in diet.

The values of SDA are: for proteins 30%, for lipids 15%, and for carbohydrates, 5%. Hence for a mixed diet, an extra 10% calories should be provided to account for the loss of energy as SDA.

Table 26.4. Calculation for energy requirement for a 55 kg person, doing moderate work

For BMR	$= 24 \times 55$ kg	$= 1320$ kcal
+ For activity	$= 40\%$ of BMR	$= 528$ kcal
Subtotal	$= 1320 + 528$	$= 1848$ kcal
+ Need for SDA	$= 1848 \times 10\%$	$= 184$ kcal
Total	$= 1848 + 184$	$= 2032$ kcal
Rounded to nearest multiple of 50		$= 2050$ kcal

Table 26.5. Fatty acids in oils

Fat or oil	Saturated (%)	Mono unsaturated (%)	Polyunsaturated (%)
Butter/ghee(*)	75	20	5
Safflower oil	9	12	79
Cotton seed oil	26	19	65
Coconut oil(*)	86	12	2
Ground nut oil	18	46	36

(*) Butter/ghee contains short chain fatty acids and coconut oil contains medium chain fatty acids

Table 26.6. Cholesterol content of food items

Food item	Cholesterol content mg/100 gm
Hens egg, whole	300
Egg yolk	1330
Liver	300-600
Brain	2000
Butter	280
Meat and fish	40-200

Table 26.7. Nitrogen balance

A normal healthy adult is said to be in nitrogen balance (Fig. 19.1), because the dietary intake (I) equals the daily loss through urine (U), feces (F) and skin (S), or, $I = U + F + S$. When the excretion exceeds intake, it is **negative** nitrogen balance. When the intake exceeds excretion, it is a state of **positive** nitrogen balance.

Factors affecting nitrogen balance

  i. **Growth:** When a person gains 5 kg, about 1 kg proteins (160 g nitrogen) are added to the body.

 ii. **Hormones:** Growth hormone, insulin and androgens promote positive nitrogen balance, while corticosteroids cause a negative nitrogen balance.

iii. **Pregnancy:** A pregnant woman will be in a state of positive nitrogen balance due to growth of fetus.

iv. **Convalescence:** A person convalescing after an illness or surgery will be in positive nitrogen balance, due to active regeneration of tissues.

 v. **Acute illness:** Negative nitrogen balance is seen immediately after surgery, trauma and burns.

vi. **Protein deficiency:** The deficiency of even a single amino acid can cause negative nitrogen balance. Starvation is another important cause.

Table 26.8. Limiting amino acids in proteins

Proteins	Limiting amino acids	Protein supplemented to cancel deficiency
Rice	Lys, Thr	Pulse proteins
Wheat	Lys, Thr	Pulse proteins
Tapioca	Phe, Tyr	Fish proteins
Bengal gram	Cys, Met	Cereals

Note: If a particular protein is fed to a young rat as the only source of protein, it fails to grow. This essential amino acid that is lacking in that protein is said to be the **limiting amino acid**. Limiting amino acid is that which limits the weight gain when a protein is supplied to an animal. This problem may be overcome by taking a mixture of proteins in the diet. **Mutual supplementation of proteins** is thus achieved. For example, pulses are deficient in methionine, but rich in lysine. On the other hand, cereals are deficient in lysine, but rich in methionine. Therefore a combination of pulses plus cereal will cancel each other's deficiency and become equivalent to first class protein.

Table 26.9. Comparison between the salient features of kwashiorkor and marasmus

	Marasmus	Kwashiorkor
Age of onset	Below one year	One to five year
Deficiency of	Calorie	Protein
Cause	Early weaning and repeated infection	Starchy diet after weaning, precipitated by an acute infection
Growth retardation	Marked	Present
Attitude	Irritable and fretful	Lethargic and apathetic
Appearance	Shrunken with skin and bones only. Dehydrated	Looks plump due to edema on face and lower limbs
Appetite	Normal	Anorexia
Skin	Dry and atrophic	'Crazy pavement dermatitis' due to peeling, cracking and denudation
Hair	No characteristic change	Sparse, soft and thin hair; curls may be lost
Associated features	Other nutritional deficiencies; Watery diarrhea Muscles are weak and atrophic	Angular stomatitis and cheilosis are common; Watery diarrhea Muscles undergo wasting
Serum albumin	2 to 3 g/dl	< 2 g/dl
Serum cortisol	Increased	Decreased

Section 8
Molecular Biology

27. Nucleotides

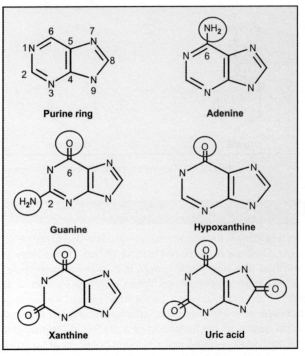

Fig. 27.1. Structure of purines

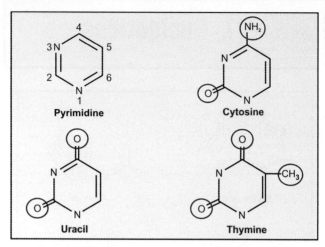

Fig. 27.2. Common pyrimidines

Table 27.1. Nucleosides and nucleotides

Nucleosides are formed when bases are attached to the pentose sugar, D-ribose or 2-deoxy D-ribose. When the nucleoside is esterified to a phosphate group, it is called a **nucleotide** or nucleoside mono-phosphate. When a second phosphate gets esterified to the existing phosphate group, a nucleoside diphosphate is generated. The attachment of a 3rd phosphate group results in the formation of a nucleoside triphosphate. The nucleic acids (DNA and RNA) are polymers of nucleoside monophosphates.

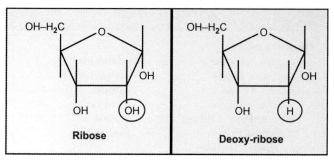

Fig. 27.3. Sugar groups in nucleic acids

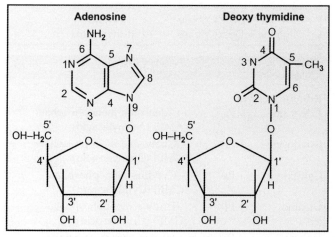

Fig. 27.4. Numbering in base and sugar groups.
Atoms in sugar is denoted with primed numbers

Table 27.2. Base + sugar = nucleosides

Ribonucleosides

Adenine + Ribose	→	Adenosine
Guanine + Ribose	→	Guanosine
Uracil + Ribose	→	Uridine
Cytosine + Ribose	→	Cytidine
Hypoxanthine + Ribose	→	Inosine
Xanthine + Ribose	→	Xanthosine

Deoxy ribonucleosides

Adenine + Deoxy ribose	→	Deoxy adenosine (d-adenosine)
Guanine + Deoxy ribose	→	d-guanosine
Cytosine + Deoxy ribose	→	d-cytidine
Thymine + Deoxy ribose	→	d-thymidine

Table 27.3. Base + sugar + phosphate = nucleotide

Ribonucleotides

Adenosine	+ Pi	→Adenosine monophosphate (AMP) (Adenylic acid)
Guanosine	+ Pi	→Guanosine monophosphate (GMP) (Guanylic acid)
Cytidine	+ Pi	→Cytidine monophosphate (CMP) (Cytidylic acid)
Uridine	+ Pi	→Uridine monophosphate (UMP) (Uridylic acid)
Inosine	+ Pi	→Inosine monohosphate (IMP) (Inosinic acid)

Table 27.4. Nucleosides and nucleotides

Base	Sugar	Nucleoside	Phosphoric acid at	Nucleotide
Adenine	Ribose	Adenosine	5' position	AMP
Do	Do	Do	3' position	3'-AMP
Do	Deoxyribose	d-adenosine	5' position	d-AMP
Do	Do	Do	3' position	d-3'-AMP
Cytosine	Ribose	Cytidine	5' position	CMP
Do	Do	Do	3' position	3'-CMP
Do	Deoxyribose	d-cytidine	5' position	d-CMP
Do	Do	Do	3' position	d-3'-CMP

Table 27.5. Nucleoside triphosphates

Nucleoside	Nucloside monophosphate	Nucleoside diphosphate (NDP)	Nucleoside triphosphate (NTP)
Adenosine	Adenosine monophosphate (AMP)	Adenosine diphosphate (ADP)	Adenosine triphosphate (ATP)
Guanosine	GMP	GDP	GTP
Inosine	IMP	IDP	ITP
Cytidine	CMP	CDP	CTP
Uridine	UMP	UDP	UTP

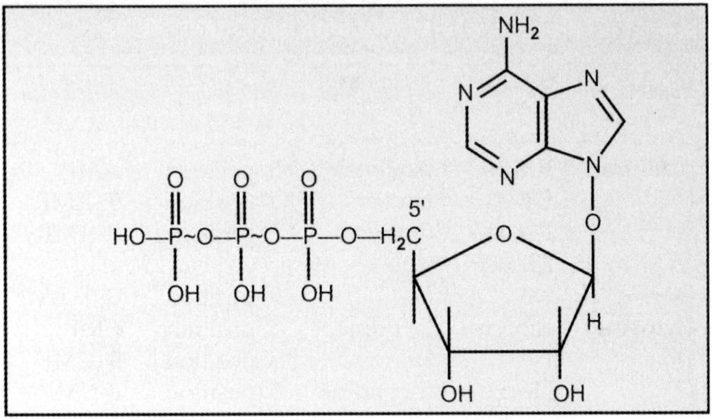

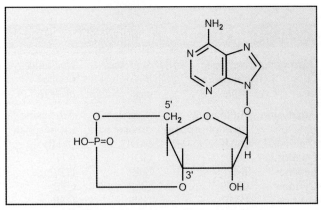

Fig. 27.5. Adenosine triphosphate (ATP)

Fig. 27.6. 3', 5'-cyclic AMP or cAMP

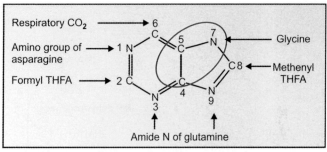

Fig. 27.7. The assembly of purine ring is from various sources.
THF4 = tetra hydro folic acid

Table 27.6. Summary of steps of de novo synthesis of purine

Steps	Donors	Added atoms
1	Glutamine	N9 (Rate-limiting)
2	Glycine	C4, 5, N7 (ATP required)
3	Methenyl-THFA	C 8
4	Glutamine	N3 (ATP required)
5	–	Ring closure (ATP)
6	Carbon dioxide	C 6
7	Aspartic acid	N1 (ATP required)
8	–	Fumarate removed
9	Formyl-THFA	C2
10	–	Ring closure IMP (Inosine monophosphate generated)

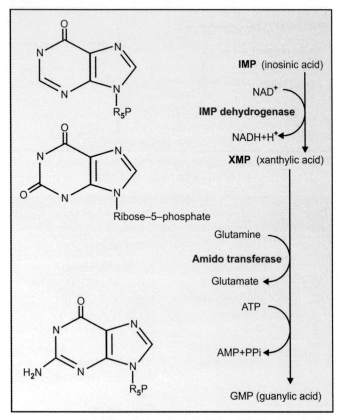

Fig. 27.8. Conversion of IMP to GMP.
R-5-P = ribose-5-phosphate

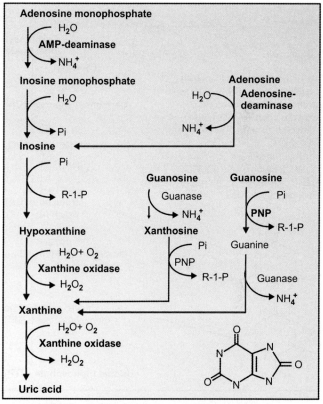

Fig. 27.9. Degradation of purine nucleotides. Main pathway is in red arrows. PNP = purine nucleoside phosphorylase; R-1-P = ribose-1-phosphate

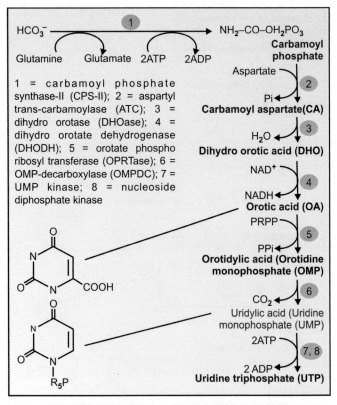

Fig. 27.10. Synthesis of pyrimidine nucleotides

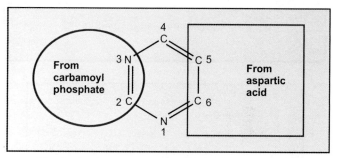

Fig. 27.11. Sources of C and N atoms of pyrimidine

Table 27.7. Gout

It is due to accumulation of urate crystals in the synovial fluid resulting in inflammation leading to acute arthritis. At 30°C, the solubility of uric acid is lowered to 4.5 mg/dl. Therefore uric acid is deposited in cooler areas of the body to cause tophi. Thus tophi are seen in distal joints of foot. Increased excretion of uric acid may cause deposition of uric acid crystals in the urinary tract leading to calculi or stone formation with renal damage. Gout may be either primary or secondary.

Gouty attacks may be precipitated by high purine diet and increased intake of alcohol. The typical gouty arthritis affects the first metatarsophalangeal joint (big toe), but other joints may also be affected. The joints are extremely painful. Synovial fluid will show urate crystals.

28. DNA Replication

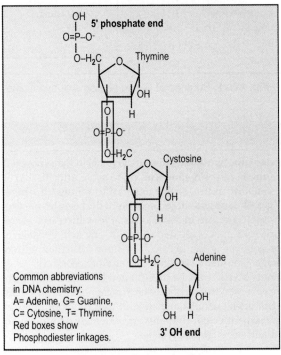

OH **5' phosphate end**
O=P–O⁻
O–H₂C ⎯ Thymine
OH
H

O=P–O⁻
O–H₂C ⎯ Cystosine
OH
H

O=P–O⁻
O–H₂C ⎯ Adenine
OH
OH   H
3' OH end

Common abbreviations
in DNA chemistry:
A= Adenine, G= Guanine,
C= Cytosine, T= Thymine.
Red boxes show
Phosphodiester linkages.

Fig. 28.1. Polynucleotide

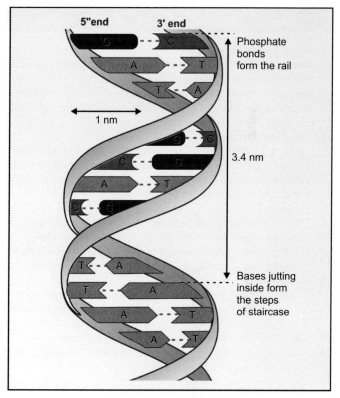

Fig. 28.2. Watson-Crick model of double helical structure of DNA

Table 28.1. Watson-Crick model of DNA structure

1. **Right handed double helix**
   DNA consists of two polydeoxy ribonucleotide chains twisted around one another in a right handed double helix similar to a spiral stair case. The sugar and phosphate groups comprise the handrail and the bases jutting inside represent the steps of the staircase.
2. **The base pairing rule**
   Always the two strands are **complementary** to each other. The base pairing (**adenine with thymine; guanine with cytosine**) is called **Chargaff's rule**, so, the number of purines is equal to the number of pyrimidines.
3. **Hydrogen bonding**
   The DNA strands are held together mainly by hydrogen bonds between the purine and pyrimidine bases. There are two hydrogen bonds between A and T while there are three hydrogen bonds between C and G.
4. **Antiparallel**
   The two strands in a DNA molecule run antiparallel. One strand runs in the 5' to 3' direction, while the other is in the 3' to 5' direction.
5. The spiral has a pitch of 3.4 nanometers per turn.
6. Within a single turn, 10 base pairs are seen. Thus, adjacent bases are separated by 0.34 nm.

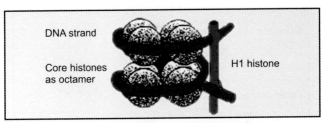

Fig. 28.3. DNA wraps twice around histone octamer to form one nucleosome

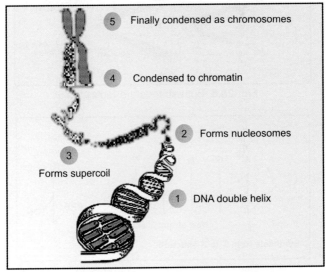

Fig. 28.4. DNA condenses repeatedly

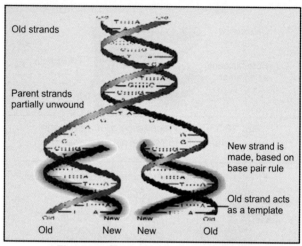

Fig. 28.5. Both strands are replicated

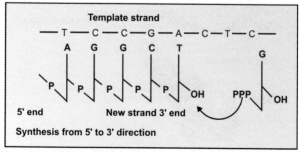

Fig. 28.6. New strand is synthesised from 5' to 3' direction. Base pairing rule is always maintained

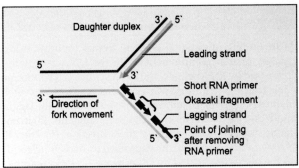

Fig. 28.7. Lagging strand and Okazaki pieces

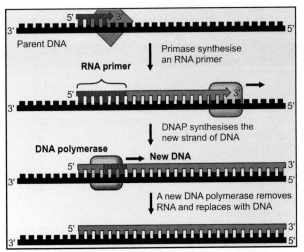

Fig. 28.8. RNA primer for the DNA synthesis

Table 28.2. Summary of DNA replication

(1) Unwinding of parental DNA to form a replication fork.
(2) RNA primer complementary to the DNA template is synthesised by RNA primase. (3) DNA synthesis is continuous in the leading strand (towards replication fork) by DNA polymerase. (4) DNA synthesis is discontinuous in the lagging strand (away from the fork), as Okazaki fragments. (5) In both strands, the synthesis is from 5' to 3' direction. (6) Then the RNA pieces are removed; the gaps filled by deoxynucleotides and the pieces are ligated by DNA ligase. (7) Proofreading is done by the DNA polymerase. (8) Finally organised into chromatin.

Table 28.3. Inhibitors of DNA replication

Drug	Action (inhibition of)
Antibacterial agents	
Ciprofloxacin	Bacterial DNA gyrase
Nalidixic acid	Do
Novobiocin	Do
Anticancer agents	
Etoposide	Human topo-isomerase
Adriamycin	Do
Doxorubicin	Do
6-mercaptopurine	Human DNA polymerase
5-fluorouracil	Do

29. Transcription and Translation

Table 29.1. Differences between RNA and DNA

RNA	DNA
1. Mainly seen in **cytoplasm**	Mostly inside **nucleus**
2. Usually 100-5000 bases	Millions of base pairs
3. Generally single stranded	Double stranded
4. Sugar is **ribose**	Sugar is **deoxyribose**
5. Purines: Adenine, guanine Pyrimidines: Cytosine, **uracil**	Adenine, guanine Cytosine, **thymine**
6. Guanine content is not equal to cytosine and adenine is not equal to uracil	Guanine is equal to cytosine and adenine is equal to thymine
7. Easily destroyed by alkali	**Alkali resistant**

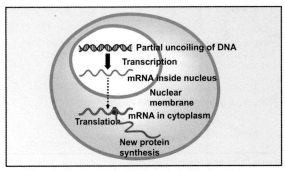

Fig. 29.1. Central dogma of molecular biology

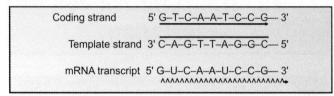

Fig. 29.2. Transcription. The mRNA base sequence is complementary to that of the template strand and identical to that of the coding strand. In mRNA, U replaces T

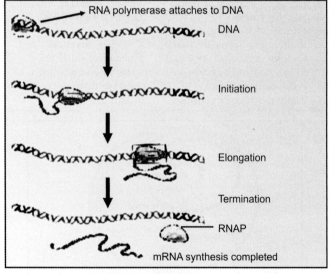

Fig. 29.3. Transcription process

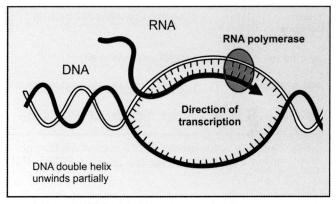

Fig. 29.4. DNA unwinds for transcription process

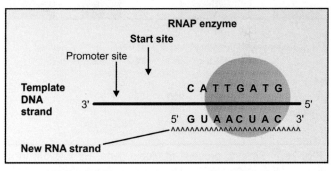

Fig. 29.5. Elongation process of transcription

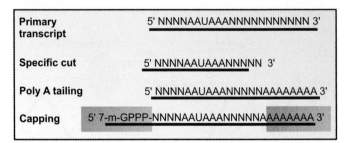

Fig. 29.6. Poly-A tail, usually 20-250 nucleotides long at 3' end. Cap at 5' terminus. N = any nucleotide

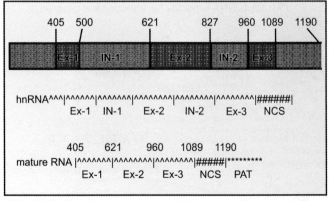

Fig. 29.7. Splicing process of globin gene. Ex = exon; In = intron; NCS = noncoding sequence (untranslated regions); PAT = poly A tail

Table 29.2. Inhibitors of RNA synthesis

Inhibitor	Source	Mode of action
Actino-mycin-D	Antibiotics from streptomyces	Insertion of phenoxa-zone ring between two G-C bp of DNA
Rifampicin	Synthetic derivative of rifamycin	Binds to beta subunit of RNA polymerase which is inactivated.
Alpha amanitin	Toxin from mushroom	Inactivates RNA polymerase II
3'-deoxy adenosine	Synthetic analogue	Incorrect entry into chain causing chain termination

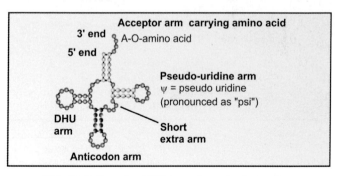

Fig. 29.8. Transfer RNA has clover-leaf structure

Table 29.3. Triplet codons and corresponding amino acids

First nucleotide 5' end	Second nucleotide				Third nucleotide 3' end
	U	C	A	G	
U	Phe	Ser	Tyr	Cys	U
	Phe	Ser	Tyr	Cys	C
	Leu	Ser	stop	stop	A
	Leu	Ser	stop	Trp	G
C	Leu	Pro	His	Arg	U
	Leu	Pro	His	Arg	C
	Leu	Pro	Gln	Arg	A
	Leu	Pro	Gln	Arg	G
A	lle	Thr	Asn	Ser	U
	lle	Thr	Asn	Ser	C
	lle	Thr	Lys	Arg	A
	Met	Thr	Lys	Arg	G
G	Val	Ala	Asp	Gly	U
	Val	Ala	Asp	Gly	C
	Val	Ala	Glu	Gly	A
	Val	Ala	Giu	Gly	G

Table 29.4. Salient features of the genetic code

i. **Triplet codons.** The codes are on the mRNA. Each codon is a consecutive sequence of three bases on the mRNA, e.g. UUU codes for phenylalanine.

ii. **Non-overlapping.** The codes are consecutive. Therefore the starting point is extremely important. The codes are read one after another in a continuous manner, e.g. AUG, CAU, CAU, GCA, etc.

iii. **Non-punctuated.** There is no punctuation between the codons. It is consecutive or continuous.

iv. **Degeneracy.** Table 29.3 shows that 61 codes stand for the 20 amino acids. So **one amino acid has more than one codon**, e.g. serine has 6 codons; while glycine has 4 codons.

v. **Unambiguous.** Though the codons are degenerate, they are unambiguous; or without any doubtful meaning. That is, one codon stands only for one amino acid.

vi. **Universal.** The codons are the same for the same amino acid in all species; the same for "Elephant and *E. coli*". The genetic code has been highly preserved during evolution.

vii. **Special codons.** UAA, UAG, and UGA are "nonsense" or "terminator" codons. They put "full stop" to the protein synthesis. AUG acts as the initiator codon.

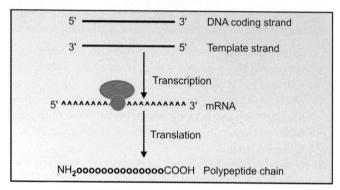

Fig. 29.9. Expression of a gene into a protein

Table 29.5. Translation process

The translation is a cytoplasmic process. The mRNA is translated from **5' to 3' end**. In the polypeptide chain synthesised, the first amino acid is the amino terminal one (Fig. 29.11). The chain growth is from amino terminal to carboxyl terminal. The process of translation can be conveniently divided into the phases of:

A. Activation of amino acid
B. Initiation
C. Elongation
D. Termination and
E. Post-translational processing.

Table 29.6. Activation of amino acid

The enzymes **aminoacyl tRNA synthetases** activate the amino acids. The enzyme is highly selective in the recognition of both the amino acid and the transfer RNA acceptor.   The CCA 3' terminus of the acceptor arm carries the amino acid.   The carboxyl group of the amino acid is esterified with 3' hydroxyl group of tRNA.

$$\text{Amino acid} + \text{tRNA} + \text{ATP} \xrightarrow{\underset{\text{tRNA synthetase}}{\text{Amino acyl}}} \text{Aminoacyl tRNA} + \text{AMP}$$

ATP is hydrolysed to AMP level, and so two high energy phosphate bonds are consumed.

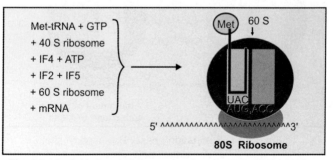

Fig. 29.10. Initiation; UAC = anticodon on met-tRNA; AUG = start signal; P = peptidyl site; A = amino acyl site

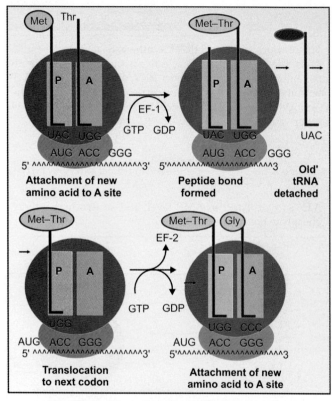

Fig. 29.11. Elongation phase. P = peptidyl site;
A = amino acyl site

Table 29.7. Termination process of translation

i. After successive addition of amino acids, ribosome reaches the **terminator codon** sequence (UAA, UAG or UGA) on the mRNA. Since there is no tRNA bearing the corresponding anticodon sequence, the "A" site remains free.

ii. The **releasing factor** (RF) enters this site along with hydrolysis of **GTP** to GDP. The RF hydrolyses the peptide chain from the tRNA at the P site.

iii. The completed peptide chain is now released. Finally 80S ribosome **dissociates** into its component units of 60S and 40S.

Table 29.8. Post-translational processing

a. Conversion of pro-insulin to **insulin** by proteolytic cleavage.

b. **Gamma carboxylation** of glutamic acid residues of prothrombin, under the influence of vitamin K.

c. **Hydroxylation** of proline and lysine in collagen with the help of vitamin C.

d. **Glycosylation**: Carbohydrates are attached to serine or threonine residues.

Table 29.9. Inhibitors of protein synthesis

The modern medical practice is heavily dependent on the use of **antibiotics**. They generally act only on bacteria and are nontoxic to human beings. This is because mammalian cells have 80S ribosomes, while bacteria have 70S ribosomes.

Reversible Inhibitors in Bacteria
These antibiotics are **bacteriostatic. Tetracyclins** bind to the ribosome and so inhibit attachment of aminoacyl tRNA to the A site of ribosomes. **Chloramphenicol** inhibits the peptidyl transferase activity of bacterial ribosomes. **Erythromycin** and **clindamycin** prevent the translocation process.

Irreversible Inhibitors in Bacteria
These antibiotics are **bactericidal. Streptomycin** causes misreading of mRNA.

Inhibitors of transcription will also in turn inhibit translation process.

Table 29.10. Mutations

Mutation may be defined as an abrupt spontaneous origin of new character.

Classification of mutations
A point mutation is defined as change in a single nucleotide. This may be subclassified as (a) substitution; (b) deletion and (c) insertion. All of them may lead to mis-sense, nonsense or frameshift effects.

Substitution
Replacement of a purine by purine (A to G or G to A) or pyrimidine by pyrimidine (T to C or C to T) is called **transition mutation**. If a purine is changed to a pyrimidine (e.g. A to C) or a pyrimidine to a purine (e.g. T to G), it is called a **transversion.** The point mutation present in DNA is transcribed and translated, so that the defective gene produces an abnormal protein.

Deletion
Large gene deletions, e.g. alpha thalassemia (entire gene) or hemophilia (partial). **Deletion of a codon**, e.g., cystic fibrosis (one amino acid, 508th phenyl alanine is missing in the CFTR protein. **Deletion of a single base**, which will give rise to frame-shift effect.

Insertion
Single base additions, leading to frame-shift effect.
Trinucleotide expansions. In Huntington's chorea, CAG trinucleotides are repeated 30 to 300 times.

Table 29.11. Effects of mutations

Silent mutation
A point mutation may change CUA to CUC; both code for leucine, so this mutation has no effect.

Mis-sense but acceptable mutation
A change in amino acid may be produced in the protein; but with no functional consequences. For example, HbA b-67 is valine (codon GUU). If a point mutation changes it to GCU, the amino acid becomes alanine; this is called **Hb Sydney**. This variant is functionally normal.

Mis-sense; partially acceptable mutation
HbS or sickle-cell hemoglobin is produced by a mutation of the beta chain in which the 6th position is changed to valine, instead of the normal glutamate. Here, the normal codon GAG is changed to GUG

Mis-sense; unacceptable mutation
HbM results from histidine to tyrosine substitution (CAU to UAU) of the distal histidine residue of alpha chain. There is met-hemoglobinemia which considerably decreases the oxygen carrying capacity of hemoglobin.

Nonsense; terminator codon mutation
A tyrosine (codon, UAC) may be mutated to a termination codon (UAA or UAG). This leads to **premature** termination of the protein, and so functional activity may be destroyed, e.g. beta-thalassemia.

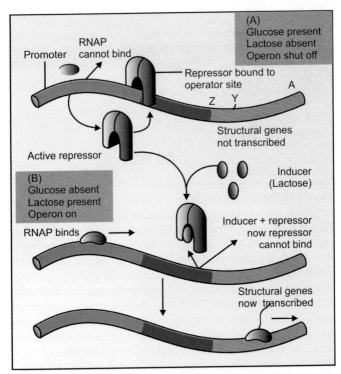

Fig. 29.12. (A) Repression of Lac operon. When lactose is absent, repressor molecules fit in the operator site. So RNAP cannot work, and genes are in "off" position. (B) Induction or Derepression of Lac operon. Lactose attaches with repressor; so operator site is free; genes are in "on" position

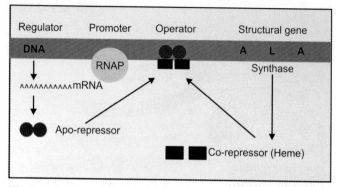

Fig. 29.13. Repression. It is the mechanism by which the presence of excess product of a pathway shuts off the synthesis of the key enzyme of that pathway. Heme synthesis is regulated by repression of **ALA synthase**, the key enzyme of the pathway. The regulatory gene produces the **apo-repressor**, which binds with heme (**co-repressor**) and becomes the active holo-repressor. The **holo-repressor** binds to the operator and stops transcription of the gene. The RNA-polymerase (RNAP) attaches at the promoter site, and starts mRNA synthesis. The operator site is in between promoter and structural genes. So when RNAP reaches operator site, it cannot move further. So enzyme synthesis stops, and heme synthesis slows down. When heme is not available, co-repressor is not available, therefore, repression is not effective and enzyme synthesis starts

30. Recombinant DNA Technology and other Molecular Biology Techniques

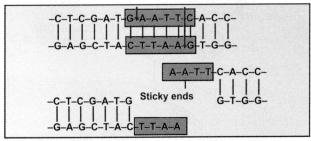

-C-T-C-G-A-T-G*A-A-T-T-C-A-C-C-
-G-A-G-C-T-A-C-T-T-A-A*G-T-G-G-

A-A-T-T-C-A-C-C-
G-T-G-G-

Sticky ends

-C-T-C-G-A-T-G
-G-A-G-C-T-A-C-T-T-A-A

Fig. 30.1. Eco RI enzyme cuts the bonds marked with red arrow. This results in the sticky ends

Table 30.1. Specificity of restriction enzymes (The arrows show the site or cut by the enzyme)

Enzyme	Source of enzyme	Specific sequence identified by enzyme		
Eco RI	*Escherichia coli* RY 13	G	AATT	C
		C	TTAA	G
Hind III	*Haemophilus influenzae* Rd	A	AGCT	T
		T	TCGA	A

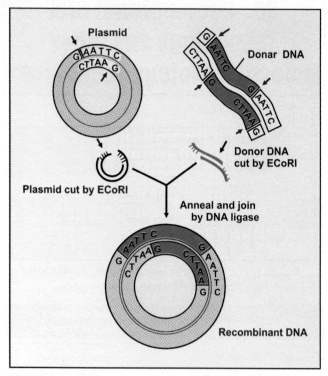

Fig. 30.2. Production of chimeric DNA molecule by using ECoRI restriction endonuclease. Red arrows show the site of cut by ECoRI

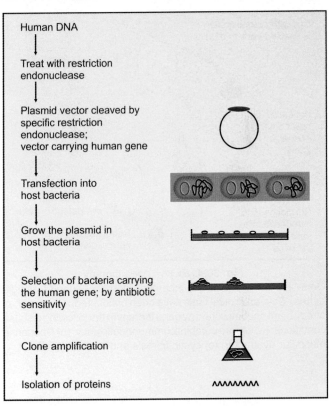

Human DNA

↓

Treat with restriction
endonuclease

↓

Plasmid vector cleaved by
specific restriction
endonuclease;
vector carrying human gene

↓

Transfection into
host bacteria

Grow the plasmid in
host bacteria

↓

Selection of bacteria carrying
the human gene; by antibiotic
sensitivity

↓

Clone amplification

↓

Isolation of proteins

Fig. 30.3. DNA-recombinant technology

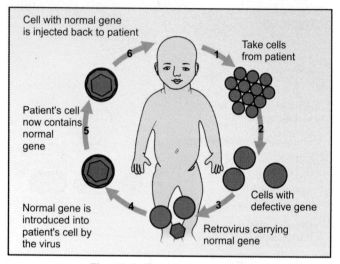

Cell with normal gene
is injected back to patient

Take cells
from patient

6

1

Patient's cell
now contains
normal
gene

5

2

Cells with
defective gene

Normal gene is
introduced into
patient's cell by
the virus

4

3

Retrovirus carrying
normal gene

Fig. 30.4. *Ex vivo* gene therapy

Gene therapy is effective in inherited disorders caused by single genes. Several clinical trials have been conducted. Success has alredy been accomplished by gene therapy in the following disease conditions: (a) severe combined immunodeficiency; (b) Duchenne muscular dystrophy; (c) cystic fibrosis and (d) hemophilia

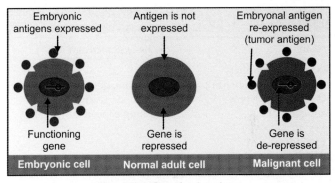

Fig. 30.5. Oncofetal antigen

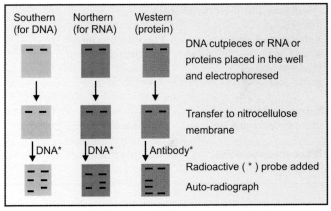

Fig. 30.6. Comparison of blot transfer techniques

Table 30.2. Important tumor markers

Name	*Serum level increased in*
Oncofetal products	
Alpha fetoprotein(AFP)	Hepatoma, germ cells cancers
Carcino embryonic antigen (CEA)	Colorectal, gastrointestinal, and lung cancer
Enzymes	
Alkaline phosphatase	Bone secondaries
Placental ALP(Regan)	Lung, seminoma
Prostate Specific Antigen (PSA)	Prostate cancer
Hormones and their metabolites	
Beta-HCG	Choriocarcinoma
Vanillyl mandelic acid (VMA)	Pheochromocytoma and neuroblastoma
Hydroxy indole acetic acid	Carcinoid syndrome
Serum proteins	
Immunoglobulins (Ig)	Multiple myeloma, macroglobulinemia
Bence-Jones proteins (in urine)	Multiple myeloma

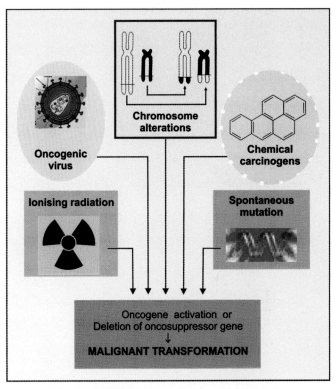

Fig. 30.7. Unified concept of carcinogenesis

Appendix I
Normal Values (Reference Values)

P = plasma; B = blood; S = serum; E = erythrocyte; U = urine;
CSF = cerebrospinal fluid; pg = picogram; ng = nanogram;
μg = microgram; mg = milligram; d = day

Analyte	Sample	Units
Alanine amino transferase		
(ALT/SGPT) Male:	S	13-35 IU/L
Female:		10-30 IU/L
Albumin	S	3.5-5 g/dl
Alkaline phosphatase (ALP)	S	40-125 IU/L
Alpha fetoprotein (AFP)	S	5-15 μg/L
Aspartate amino transferase		
(AST/SGOT)	S	8-20 IU/L
Bicarbonate (HCO_3^-)	S	22-26 mEq/L
Bilirubin, total	S	0.2-1 mg/dl
Calcium	S	9-11 mg/dl
Ceruloplasmin	S	25-50 mg/dl
Chloride	S/P	96-106 mEq/L
Chloride	CSF	120-130 mEq/L
Cholesterol, total	S/P	150-200 mg/dl

Contd...

Contd...

Analyte	Sample	Units
(HDL fraction) Male	S	30-60 mg/dl
Female		35-75 mg/dl
(LDL fraction) 20-29 yr		60-150 mg/dl
30-39 yr		80-175 mg/dl
Creatine	S	0.2-0.4 mg/dl
Creatine kinase (CK)		
Female	S	10-80 U/L
Male	S	15-100 U/L
Creatinine	S	0.7-1.4 mg/dl
	U	15-25 mg/kg/d
Fibrinogen	P	200-400 mg/dl
Globulins	S	2.5–3.5 g/dl
Glucose (Fasting)	P	70-110 mg/dl
	CSF	50-70 mg/dl
Hemoglobin Male	B	14-16 g/dl
Female	B	13-15 g/dl
Hb A₁c (glycohemoglobin)		4-8% of total
Immunoglobulins	S	
IgG		800-1200 mg/dl
IgM		50-200 mg/dl
IgA		150-300 mg/dl
Iron	S	100-150 μg/dl
Iron binding capacity	S	250-400 μg/dl
Lactate dehydrogenase	S	100-200 IU/L
Lipids—total	S	400-600 mg/dl

Contd...

Analyte	Sample	Units
Lipoproteins Alpha	S	40 mg/dl
Beta		180 mg/dl
Nonesterified fatty acids	P	10-20 mg/dl
Parathyroid hormone	S	10-25 ng/L
pCO_2, arterial	B	35-45 mm Hg
pH	B	7.4
Phosphate	S	3-4 mg/dl
	U	1 g/day
Phospholipids		150-200 mg/dl
pO_2 arterial	B	90-100 mm Hg
Potassium	S	3.5-5 mEq/L
Proteins—total	S	6-8 g/dl
Prothrombin	P	10-15 mg/dl
Sodium	S	136-145 mEq/L
T_3 (Tri iodothyronine)	S	120-190 ng/dl
T_4 (thyroxine)	S	5-12 µg/dl
TSH	S	0.5-5 µU/mL
Triglycerides, fasting, Male,	S	50-200 mg/dl
Female		40-150 mg/dl
Urea	S	20-40 mg/dl
Uric acid, Male	S/P	3.5-7 mg/dl
Female	S/P	3.0-6 mg/dl
Vitamin A	S	15-50 µg/dl
Vitamin C (Ascorbic acid)	P	0.4-1.5 mg/dl
Vitamin D3 (Calcitriol)	S	1.5-6 µg/dl
Vitamin E	S	0.5-1.8 mg/dl

Appendix II
Composition of Nutrients in Common Food Materials

Food materials	Protein g/100 g	Fat g/100 g	Carbo-hydrate g/100 ml
I. Cereals:			
1. Wheat flour, whole	12.1	1.7	72.2
2. Rice, raw, milled	6.9	0.4	79.2
II. Legumes and pulses:			
1. Bengal gram, (Channa)	17.1	5.3	61.2
2. Peas (Mattar) dried	19.7	1.1	56.6
3. Soyabean	43.2	19.5	20.9
III. Vegetables, A group, (low calorie)			
1. Amaranth (lal cholai)	4.9	0.5	5.7
2. Cabbage	1.8	0.1	6.3
3. Tomato, ripe	1.0	0.1	3.9
IV. Vegetables, B group, (Medium calory)			
1. Carrot	0.9	0.2	10.7
2. Onion, big (sabola)	1.2	—	11.6
V. Vegetables, C group (Roots and Tubers)			
1. Potato (Aloo)	1.6	0.1	22.9
2. Tapioca (cassava)	0.7	0.2	38.7
3. Yam (Ratalu)	1.4	0.1	27.0

Contd...

Contd...

Food materials	Protein g/100 g	Fat g/100 g	Carbo-hydrate g/100 ml
VI. Fruits			
1. Apple	0.3	0.1	13.4
2. Banana, ripe	1.3	0.2	36.4
3. Mango, ripe	0.6	0.1	11.8
4. Papaya, ripe	0.5	0.1	9.5
VII. Milk and milk products			
1. Cow's milk	3.3	3.8	4.4
2. Buffalo's milk	4.3	8.8	5.3
3. Curd (Yogurt) (dahi)	2.9	2.9	3.3
4. Cheese (Paneer)	24.1	25.1	6.3
VIII. Meat and other products			
1. Mutton, muscle	18.5	11.3	0.5
2. Beef muscle	22.6	2.6	0.5
3. Fish	22.6	0.6	0.2
4. Egg, Hen	13.3	13.3	0.2

Food materials	Calcium mg/ 100 g	Iron mg/ 100 g	Vit. A IU/ 100 g	Vit. B$_1$ µg/ 100 g
I. Cereals:				
1. Wheat flour, whole	35	7.3	—	70
2. Rice, raw, milled	10	1.0	—	50
3. Sorghum, Juar, Cholam	30	6.2	136	345
II. Legumes and Pulses:				
1. Bengal gram, (Channa)	190	9.8	316	300
2. Peas (Mattar) dried	70	4.4	—	450
3. Soyabean	240	11.5	710	730
III. Vegetables, A group, (low calorie)				
1. Amaranth (lal cholai)	500	21.4	8,000	50
2. Cabbage	30	0.8	2,000	60
3. Tomato, ripe	10	—	320	120
IV. Vegetables, B group, (Medium calory)				
1. Carrot	80	1.5	4,000	40
2. Onion, big (sabola)	180	0.7	25	80
V. Vegetables, C group (Roots and Tubers)				
1. Potato (Aloo)	—	0.7	40	100
2. Tapioca (cassava)	50	0.9	—	45
3. Yam (Ratalu)	60	1.3	—	72

Contd...

Contd...

Food materials	Calcium mg/ 100 g	Iron mg/ 100 g	Vit. A IU/ 100 g	Vit. B₁ µg/ 100 g
VI. Fruits				
1. Apple	—	1.7	—	120
2. Banana, ripe	10	—	—	150
3. Mango, ripe	10	—	4,800	40
4. Papaya, ripe	10	—	3,000	40
VII. Milk and milk products				
1. Cow's milk	100	—	160	50
2. Buffalo's milk	210	—	160	40
3. Curd (Yogurt) (dahi)	120	—	130	—
4. Cheese (Paneer)	790	2.1	275	—
VIII. Meat and other products				
1. Mutton, muscle	150	2.5	30	180
2. Beef muscle	10	0.8	60	150
3. Fish	20	0.9	20	100
4. Egg, hen	60	2.1	1,200	130

Appendix III
Recommended Daily Allowance (RDA) of Essential Nutrients

Nutrient		Requirement per day
1. Proteins		
Adult	-	1 g/kg
Infants	-	2.4 g/kg
Upto 10 years	-	1.75 g/kg
Pregnancy	-	2 g/kg
2. Essential amino acids		
Phenylalanine	-	14 mg/kg
Leucine	-	11 mg/kg
Lysine	-	9 mg/kg
Valine	-	14 mg/kg
Isoleucine	-	10 mg/kg
Threonine	-	6 mg/kg
Methionine	-	14 mg/kg
Tryptophan	-	3 mg/kg
3. Fat soluble vitamins		
Vitamin A, Adult	-	750 µg
Children	-	400 to 600 µg
Pregnancy	-	1000 µg
Vitamin D, Adult	-	5 µg
Children (preschool)	-	10 µg
Pregnancy and lactation	-	1200 µg
Vitamin E	-	10 mg

Contd...

Contd...

	Nutrient		Requirement per day
	Vitamin K, Adult	-	50 to 100 mg
	Children	-	1 µg/kg
4.	**Water soluble vitamins**		
	Thiamine (B₁)	-	1-1.5 mg
	Riboflavin (B₂)	-	1.5 mg
	Niacin	-	20 mg
	Pyridoxine (B₆)	-	2 mg
	Pantothenic acid	-	10 mg
	Biotin	-	200-300 µg
	Folic acid, Adult	-	100 µg
	Pregnancy	-	300 µg
	Vitamin B₁₂	-	1 µg
	Ascorbic acid, Adult	-	70 mg
	Pregnancy and lactation	-	100 mg
5.	**Minerals**		
	Calcium, Adult	-	0.5 g
	Children	-	1 g
	Pregnancy and lactation	-	1.5 g
	Phosphorus	-	500 mg
	Magnesium	-	400 mg
	Manganese	-	5-6 mg
	Sodium	-	5-10 g
	Potassium	-	3-4 g
	Iron, Males	-	15-20 mg
	Females	-	20-25 mg
	Pregnancy	-	40-50 mg
	Copper	-	1.5-3 mg
	Iodine	-	150-200 µg
	Zinc	-	8-10 mg
	Selenium	-	50-100 µg

Index